NORTH-HOLLAND
MATHEMATICS STUDIES 107

Lecture Notes in Numerical and Applied Analysis Vol. 7
General Editors: H. Fujita and M. Yamaguti

Foundations of the Numerical Analysis of Plasticity

TETSUHIKO MIYOSHI (Yamaguchi University)

KINOKUNIYA
KINOKUNIYA COMPANY LTD.
TOKYO JAPAN

NORTH-HOLLAND PUBLISHING COMPANY
AMSTERDAM · NEW YORK · OXFORD

KINOKUNIYA COMPANY - TOKYO
NORTH-HOLLAND PUBLISHING COMPANY - AMSTERDAM · NEW YORK · OXFORD

ISBN: 0 444 87671 5

Publishers
KINOKUNIYA COMPANY LTD.
TOKYO JAPAN

* * *

ELSEVIER SCIENCE PUBLISHERS
B. V. (NORTH-HOLLAND)
AMSTERDAM • OXFORD • NEW YORK

Sole distributors for the U.S.A. and Canada
ELSEVIER SCIENCE PUBLISHING COMPANY. INC.
52 VANDERBII.T AVENUE
NEW YORK. N.Y. 10017

Distributed in Japan by KINOKUNIYA COMPANY LTD.
Distributed outside Japan by ELSEVIER SCIENCE PUBLISHERS B. V. (NORTH-HOLLAND)

Lecture Notes in Numerical and Applied Analysis Vol. 7

PRINTED IN JAPAN

PREFACE

In the past several decades the techniques to solve partial differential equations numerically have made remarkable progress, and various nonlinear problems in mathematical physics have come within the range of numerical analysis. The analysis of plasticity is a typical case in structural mechanics. By applying the finite element techniques we can get the numerical solutions of the initial-boundary-value problems which express the complicated deformation process of plastic bodies. This progress, of course, is supported by the development of the highspeed electric computer. And now we have extensive computational results on plasticity.

The plasticity problem is nonlinear in the usual sense that the differential operator is algebraically nonlinear. However, this problem has another nonlinearity due to the hysteresis character of the material, and this causes various difficulties for both mathematical and numerical treatments. In fact, the existing numerical algorithms require an enormous amount of calculation to solve actual problems in engineering ,and even the error estimate or the convergence proof is still not completely given for basic algorithms.

The main purpose of this monograph is to describe a theoretical foundation for analysing and developing approximate methods to solve dynamic and quasi-static plasticity problems. The plasticity theory considered here is an incremental theory supposing the existence of a plastic potential. Also, as the hardening rule, we assume kinematic and isotropic hardenings throughout this monograph.

We first treat the simplest plasticity problem in Chapter 2 and proceed to more complicated cases. A considerable part of several chapters is devoted to the mechanism that causes yielding and unloading. This is because the first question in the mathematical theory of plasticity is why the yielding or the unloading occurs , or what mechanism causes such phenomena. Once this question is answered for simple mass-spring systems, then it is not difficult to extend the results to more complicated problems. In fact, the continuous bodies can be regarded as a limit of discrete systems - finite elements, for instance. This approach is constructive, and it also enable us to discuss things within the framework of plasticity theory.

As regards the numerical methods, we placed special emphasis on the analysis of explicit integration schemes , since the explicit approximation is one of the basic methods and widely used to solve actual problems in engineering.

This monograph contains ten chapters and two appendices. In Chapter 1 we introduce the mathematical models treated in this monograph. In Chapters 2, 3 and 4 we consider spring-mass systems with elastic-plastic constitutive law. In these chapters we observe some basic mathematical characters of plastic deformation, especially the mechanism that causes yielding and unloading. In Chapters 5 and 6 we treat the dynamic and quasi-static semi-discrete systems which are derived by applying the finite element method to two-dimensional plasticity problems, and we extend the results obtained for the spring-mass systems. Chapters 7 and 8 contain the analysis of two typical computational methods of the explicit type. In Chapter 9 the convergence and error estimates of the finite element solutions are discussed. At the same time, the results of this chapter present a constructive method to prove the existence of the solutions to the original (fully continuous) problems. Chapter 10 is a trial which attempts to extend this idea of existence proof to problems with both material and geometrical nonlinearities.

This monograph was written in Kumamoto, where I worked thirteen years as one of the teaching staff at Kumamoto University. I wish to thank all my former collegues of the Department of Mathematics at Kumamoto University for their continuous support and encouragement. I am especially indebted to Dr. Alan David Rosen of Kumamoto University for helpful comments in correcting the writing of the manuscript.

Tetsuhiko Miyoshi

Yamaguchi, Japan

August, 1984

TABLE OF CONTENTS

Lecture Notes in Num. Appl. Anal., **7,** 1–249 (1984)
Foundations of the Numerical Analysis of Plasticity, 1984

CHAPTER 1

MATHEMATICAL MODELS OF ELASTIC-PLASTIC PROBLEMS

1.1 Spring-mass system with one degree of freedom

We begin with the simplest model of the elastic-plastic vibration. As is well known, the simplest model of the elastic vibration of a spring-mass system with one degree of freedom (we also call it a single mass system, for simplicity's sake) is given by the following differential equations :

$$(1.1) \qquad \rho\ddot{u} + \sigma = b \qquad\qquad \sigma = ku,$$

where ρ, u, σ, b and k are the mass, displacement, stress, outer force and elastic constant, respectively. The dot denotes the differentiation with respect to time t. We assume throughout this book that the given forces, b in the present case, are continuous and piecewise analytic with respect to time t in an interval including I = (0,T).

Now the relation between the stress and strain (σ and u in the present problem) is one to one in the elastic state, as is seen in (1.1). However, this correspondence does not usually hold beyond a certain threshold of the stress ($|\sigma| = z_0$, for example). In the theory of plasticity , the strain after this threshold is assumed to be the sum of the elastic and plastic components . It is also assumed that the former can be recovered but that the latter remains as the eternal strain.

In our case, u is divided as

(1.2) $$u = u^e + u^p,$$

or incrementally as

(1.3) $$\dot{u} = \dot{u}^e + \dot{u}^p.$$

Here $\dot{u}^e$ and $\dot{\sigma}$ preserve the relation

(1.4) $$\dot{\sigma} = k\dot{u}^e,$$

but $\dot{u}^p$ and $\dot{\sigma}$ are connected by another equation ,

(1.5) $$\dot{\sigma} = \eta\dot{u}^p \qquad (\ \eta > 0\ ),$$

where η is a certain function of the deformation history. From (1.3) and (1.5) we have

$$\frac{1}{k}\dot{\sigma} = \dot{u}^e = \dot{u} - \dot{u}^p = \dot{u} - \frac{1}{\eta}\dot{\sigma}.$$

Hence, by solving this equation with respect to $\dot{\sigma}$, we have

(1.6) $$\dot{\sigma} = \frac{\eta k}{\eta + k}\ \dot{u} = (\ k - \frac{k^2}{\eta + k}\ )\dot{u}.$$

This is the stress-strain relation in the plastic state.

We shall say that the spring-mass system, or the stress - strain relation is elastic (or plastic) when the relation $\dot{\sigma} = k\dot{u}$ (or (1.6)) is used. The transition from the elastic state to the plastic state is called the " yielding." The function η ($= d\sigma/du^p$) is called the " strain hardening " rate and depends on the history of the deformation. To determine this function we need some assumption on the hardening. However, for the time being, we assume that this is a given function. Now assume that we got a plastic solution which is continuated from the elastic one. Since the elastic limit is attained by the increasing $|u|$, $|u|$ is still increasing during a certain time interval. However, there must be such t_0 that $\dot{u}(t_0) = 0$. If $|u|$ begins to increase after t_0, then there is no question. If, however, $|u|$ is decreasing under the same stress - strain relation, then this solution is not admissible. In this case the elastic stress-strain relation must be used instead of (1.6). It can then be proved that the solution u obtained under the elastic stress-strain relation is decreasing as desired. The elastic state which follows the plastic state is the process of unloading at first.

Assume that the unloading began and the elastic deformation continued. Therefore, it may yield again after deforming in the opposite direction. Hence, we need a criterion to determine the second yielding and the next σ - u relation. Here arises again the problem of how to formulate the hardening. Let us consider two typical models of hardening.

Kinematic hardening : In this model the yield surface (this " surface " consists of two points in the present problem) translates in the σ - space without changing shape during the plastic deformation. Let α be the parameter representing the center of this surface. Then α must satisfy

$$(1.7) \qquad \begin{cases} \dot{\alpha} = 0 & \text{in the elastic state} \\ \dot{\alpha} = \dot{\sigma} & \text{in the plastic state.} \end{cases}$$

The yield surface at time t is expressed as

$$(1.8) \qquad |\sigma(t) - \alpha(t)| = z_0.$$

We always assume in this monograph that the strain hardening rate η in the kinematic hardening model is constant. Hence, by setting

$$\xi = \frac{k}{\eta + k}$$

our problem is to seek u or (u, σ, α) which satisfies the following system of equations.

(1.9) $$\rho\ddot{u} + \sigma = b \qquad \text{in } I$$

(1.10) $$\left\{\begin{array}{llll} \dot{\sigma} = k\dot{u} & \dot{\alpha} = 0 & \text{if } |\sigma - \alpha| < z_0 & \\ \dot{\sigma} = (1 - \xi)k\dot{u} & \dot{\alpha} = \dot{\sigma} & \text{if } |\sigma - \alpha| = z_0 & \text{and } (\sigma - \alpha)\dot{\sigma} \geq 0. \end{array}\right.$$

The mass system is elastic if $|\sigma - \alpha| < z_0$ and plastic if $|\sigma - \alpha| = z_0$.

REMARK. The left and right derivatives of σ and α do not necessarily coincide when the yielding occurs. In the case of the unloading, however, $\dot{\sigma}$ and $\dot{\alpha}$ are continuous for dynamic problems. See Chapter 2 for a more exact setting of the problem.

Isotropic hardening : In this model the yield surface expands monotonically. The typical formulation is to assume that, during the plastic deformation, σ satisfies

(1.11) $$|\sigma| = F(w^p),$$

where w^p is the plastic work defined by

(1.12) $$w^p = \int_0^t \sigma\,\dot{u}^p\,dt,$$

and F is a monotonically and strictly increasing function with suitable smoothness. We shall introduce below a way to determine F practically and at the same time to derive η in (1.5).

By (1.5) and the hardening assumption that the yield surface expands, $\sigma\dot{u}^p$ must be nonnegative. Hence we have by (1.12)

$$\dot{w}^p = \sigma\,\dot{u}^p = |\sigma|\;|\dot{u}^p|.$$

Since $\dot{u}^p = 0$ in the elastic state, we have

$$|\sigma| = F(w^p) = F\left(\int_0^t |\sigma||\dot{u}^p| dt \right).$$

We first prove that there is a monotone function H such that the relation

$$|\sigma| = H\left(\int_0^t |\dot{u}^p| dt \right) \tag{1.13}$$

holds in the plastic state. Since F is monotone, the function G defined by

$$G(|\sigma|)(t) = F^{-1}(|\sigma|)(t) = \int_0^t |\sigma||\dot{u}^p| dt,$$

that is, $G = F^{-1}$ is again monotone with respect to $|\sigma|$. By differentiating both sides of the above identity we have

$$G'(|\sigma|)|\sigma|' = |\sigma||\dot{u}^p| \quad \text{i.e.} \quad |\dot{u}^p| = \frac{G'(|\sigma|)}{|\sigma|}|\sigma|'$$

in the plastic state. Since we are interested in σ in the plastic state, we can assume that $|\sigma|$ is constant during the elastic deformation. Under this modification we have

$$\int_0^t |\dot{u}^p| dt = \int_0^t \frac{G'(|\sigma|)|\sigma|'}{|\sigma|} dt = \int_{z_0}^{|\sigma|} \frac{G'(|\sigma|)}{|\sigma|} d(|\sigma|).$$

Let the last term be $Q(|\sigma|)$. Q is then monotonically increasing. By setting $H = Q^{-1}$ we have the relation (1.13). Now conversely, the function F is determined if H is defined. To see this, put $Q = H^{-1}$. Then we have

$$Q(|\sigma|) = \int_0^t |\dot{u}^p| dt \quad \text{i.e.} \quad Q'(|\sigma|)|\sigma|' = |\dot{u}^p|.$$

The function G defined by

$$G(|\sigma|) = \int_{z_0}^{|\sigma|} Q'(|\sigma|)|\sigma| \, d(|\sigma|)$$

is monotonically increasing with respect to $|\sigma|$ and

$$G(|\sigma|) = \int_0^t |\sigma||\dot{u}^p| dt = \int_0^t \dot{w}^p \; dt.$$

Hence $F = G^{-1}$ is the desired function.

There are some methods to determine H in practice. A typical one is to assume that

$$H(\bar{u}^p) = c(d + \bar{u}^p)^n$$

for small $\bar{u}^p$ and linear for large $\bar{u}^p$, where

$$\bar{u}^p = \int_0^t |\dot{u}^p| dt.$$

The constants c, d and n are determined by the experiment for one deformation process, for instance, a uniaxial tensile test of the material.

Now if H is defined as a function of $\bar{u}^p$, then the function η in (1.5) is given by

(1.14) $$\eta = H'(\bar{u}^p).$$

To see this, differentiate both sides of (1.13) to get

$$\frac{d}{dt} |\sigma| = H' \; |\dot{u}^p|.$$

Since $d|\sigma|/dt = \sigma/|\sigma| d\sigma/dt$, we have by (1.5)

$$\dot{u}^p = \frac{1}{\eta}\dot{\sigma} = \frac{1}{\eta}\frac{\sigma}{|\sigma|}H'|\dot{u}^p|$$

and hence $\dot{w}^p = \sigma\dot{u}^p = 1/\eta|\sigma|H'|\dot{u}^p|$. On the other hand, since $\sigma\dot{u}^p \geq 0$, the right side of this identity must be equal to $|\sigma||\dot{u}^p|$. Hence we have $1/\eta \; H' = 1$. The governing equations under the isotropic hardening rule are written as follows. Let $H' = H'(\bar{u}^p)$ and $\bar{u}^p = \int_0^t |\dot{u}^p| \; dt$.

$$(1.15)\quad \begin{cases} \rho\ddot{u} + \sigma = b & \text{in } I \\ \dot{\sigma} = k\dot{u}, \quad \dot{u}^p = 0 & \text{if } |\sigma| < H(\bar{u}^p) \\ \dot{\sigma} = \left(k - \frac{k^2}{H'+k} \right)\dot{u}, \quad \dot{u}^p = \frac{\dot{\sigma}}{H'} & \text{if } |\sigma| = H(\bar{u}^p) \text{ and } \sigma\dot{\sigma} \geq 0. \end{cases}$$

REMARK: Assume that the mass-system is plastic for $t > t_0$. Then $\sigma\dot{\sigma} \geq 0$ and the following equations must be satisfied for $t > t_0$.

$$(1.16)\quad \begin{cases} \rho\ddot{u} + \sigma = b & \text{in } I \\ \dot{\sigma} = \left(k - \frac{k^2}{H'+k} \right)\dot{u} \\ \dot{u}^p = \frac{\dot{\sigma}}{H'} . \end{cases}$$

The initial values of u, $\dot{u}$, σ and u^p are given, of course, at $t = t_0$. Let (u, σ, u^p) be the solution of this system of equations. Then the condition $|\sigma| = H(\bar{u}^p)$ is satisfied automatically as far as $\sigma\dot{\sigma} \geq 0$ is satisfied. Hence we introduce a function g defined by

$$g(\sigma) = H'(H^{-1}(|\sigma|))$$

and consider the problem.

$$(1.17)\quad \begin{cases} \rho\ddot{u} + \sigma = b \\ \dot{\sigma} = \left(k - \frac{k^2}{g(\sigma)+k} \right)\dot{u} \end{cases}$$

with the same initial conditions as before for u, $\dot{u}$ and σ at $t = t_0$. For the solution of this system, we define $\bar{u}^p$ by

$$\bar{u}^p = \int_0^t |\dot{u}^p|\,dt, \qquad \dot{u}^p = \frac{\dot{\sigma}}{g(\sigma)} .$$

Then the identity $|\sigma| = H(\bar{u}^p)$ is again automatically satisfied as far as the condition $\sigma\dot{\sigma} \geq 0$ is satisfied. To prove this, we first note that, since $\sigma\dot{\sigma} \geq 0$ implies $|\dot{\sigma}| = d|\sigma|/dt$, we have

$$\frac{d}{dt}\bar{u}^p = |\dot{u}^p| = \frac{1}{g(\sigma)}|\dot{\sigma}| = \frac{1}{g(\sigma)}\frac{d|\sigma|}{dt},$$

so that $H'(H^{-1}(|\sigma|))\frac{d}{dt}\bar{u}^p = \frac{d|\sigma|}{dt}$. We next define $z = H^{-1}(|\sigma|)$, that is, $|\sigma| = H(z)$. Then, since

$$\frac{d|\sigma|}{dt} = H'(z)\frac{dz}{dt} = H'(H^{-1}(|\sigma|))\frac{d}{dt}H^{-1}(|\sigma|),$$

we have the following identity.

$$\frac{d}{dt}(H^{-1}(|\sigma|)) = \frac{d}{dt}\bar{u}^p.$$

Integrating both sides from t_0 to t and using the condition $|\sigma|(t_0) = H(\bar{u}^p)(t_0)$, we have the desired equality.

This observation implies that the system (1.15) is equivalent to the following system.

$$(1.18)\quad \begin{cases} \rho\ddot{u} + \sigma = b & \text{in } I \\ \dot{\sigma} = k\dot{u},\quad \dot{u}^p = 0 & \text{if } |\sigma| < H(\bar{u}^p) \\ \dot{\sigma} = \left(k - \dfrac{k^2}{g(\sigma)+k}\right)\dot{u},\quad \dot{u}^p = \dfrac{1}{g(\sigma)}\dot{\sigma} & \text{if } |\sigma| = H(\bar{u}^p) \text{ and } \sigma\dot{\sigma} \geq 0, \end{cases}$$

where

$$g(\sigma) = H'(H^{-1}(|\sigma|)),\qquad \bar{u}^p = \int_0^t |\dot{u}^p|\,dt.$$

In the sequel we use this formulation for the isotropic hardening case. Also, we assume the following condition for the function H.

ASSUMPTION : The function H belongs to the class $C^2(-\delta,\infty)$ $(\delta > 0)$ and is piecewise analytic. Also it satisfies the following conditions.

$$H(0) = z_0, \quad H' \geq \delta_0 > 0, \quad H'' < 0.$$

1.2. Spring-mass system with multiple degrees of freedom

Consider a system of N springs and masses. We also call this system a " multiple mass system." We shall derive the governing equation of this system for both kinematic and isotropic hardening rules.

Let ρ_i and u_i($i = 1,2,\ldots, N$) be the mass and displacement of the i-th point. Let k_i be the elasticity constant of the i-th spring. The elastic vibration of this system is described by the initial value problem of the following system of equations.

$$\rho_i \ddot{u}_i + \sigma_i - \sigma_{i+1} = b_i \qquad i = 1 \sim N, \tag{2.1}$$

$$\sigma_i = k_i U_i = u_i - u_{i-1} \tag{2.2}$$

As the boundary condition we give $u_0 = 0$ and $\sigma_{N+1} = 0$.

Let the initial yielding of the i-th spring be given by the condition

$$|\sigma_i| = z_0.$$

As the stress-strain relation and the hardening rule after the initial yielding we consider the following models.

1. The kinematic hardening :

$$(2.3) \quad \begin{cases} \dot{\sigma}_i = k_i \dot{U}_i, \quad \dot{\alpha}_i = 0 & \text{if } |\sigma_i - \alpha_i| < z_0 \\ \dot{\sigma}_i = (1 - \xi_i) k_i \dot{U}_i, \ \dot{\alpha}_i = \dot{\sigma}_i & \text{if } |\sigma_i - \alpha_i| = z_0 \text{ and } (\sigma_i - \alpha_i)\dot{\sigma}_i \geq 0. \end{cases}$$

2. The isotropic hardening :

$$(2.4)\quad \begin{cases} \dot{\sigma}_i = k_i\dot{U}_i, \quad \dot{U}_i^p = 0 & \text{if } |\sigma_i| < H(\bar{U}_i^p) \\ \dot{\sigma}_i = \left(k_i - \dfrac{k_i^2}{g(\sigma_i)+k_i} \right)\dot{U}_i, \quad \dot{U}_i^p = \dfrac{\dot{\sigma}_i}{g(\sigma_i)} & \text{if } |\sigma_i| = H(\bar{U}_i^p) \text{ and } \sigma_i\dot{\sigma}_i \geq 0, \end{cases}$$

where $g(\sigma_i) = H'(H^{-1}(|\sigma_i|))$ and $\bar{U}_i^p = \int_0^t |\dot{U}_i^p|\,dt$.

For the quasi-static problem, the inertia term of (2.1) is of course absent. Also, in this case the continuity of $\dot{u}$ can not be expected in general, so that we have to regard the derivative as one-sided depending on the situation. Further, it may occur that the elastic stress-strain relation must be employed on the yield surface. Hence the elastic stress - strain relation in the quasi-static problem takes the following form.

(Kinematic hardening) : $\dot{\sigma}_i = k_i\dot{U}_i, \quad \dot{\alpha}_i = 0,$

$$\text{if} |\sigma_i - \alpha_i| < z_0 \text{ or } |\sigma_i - \alpha_i| = z_0 \text{ and } (\sigma_i - \alpha_i)\dot{\sigma} < 0.$$

(Isotropic hardening) : $\dot{\sigma}_i = k_i\dot{U}_i, \quad \dot{U}_i^p = 0,$

$$\text{if } |\sigma_i| < H(\bar{U}_i^p) \text{ or } |\sigma_i| = H(\bar{U}_i^p) \text{ and } \sigma_i\dot{\sigma}_i \geq 0.$$

We consider the following boundary conditions. In the dynamic problem one end is fixed ($u_0 = 0$) and another is free ($\sigma_{N+1} = 0.$) In the quasi-static problem the both ends are fixed ($u_0 = u_{N+1} = 0.$). This is because the quasi-static problem for the mass system becomes trivial mathematically for the former boundary condition.

1.3. Two-dimensional problems

Let Ω be a domain in (x_1, x_2) plane and Γ its boundary. Let u, σ and ε be the displacement, stress and strain vectors with the following components :

$$u^* = (u_1, u_2) \qquad \sigma^* = (\sigma_{11}, \sigma_{22}, \sigma_{12}) \qquad \varepsilon^* = (\varepsilon_{11}, \varepsilon_{22}, \varepsilon_{12})$$

The notation * denotes the transposition of a vector. Thus the inner product of vectors are frequently denoted by a*b as well as (a, b). The inner product and norm of (vector) functions in $L^2(\Omega)$ are also denoted by (a,b) and $\|a\|$, respectively. Thus we use the same notation to denote the inner product and norm of vectors, single functions, and vector functions, whenever no ambiguity occurs.

Let b_i (i = 1, 2) and ρ (= constant) be the body forces and mass density of the material. The two-dimensional problem considered in this section is to find u (and σ) satisfying the following conditions.

(a) The equiliblium equations (or the equations of motion) :

$$(\rho \ddot{u}_i) - \sum_j \sigma_{ij,j} = b_i \qquad \text{in } I \times \Omega \quad (i = 1,2)$$

where $\sigma_{21} = \sigma_{12}$.

(b) The strain-displacement relation :

$$\varepsilon_{11} = u_{1,1} \qquad \varepsilon_{22} = u_{2,2} \qquad \varepsilon_{12} = u_{2,1} + u_{1,2}$$

where $u_{,j} = \partial u/\partial x_j$ (j = 1,2). Note that the strain is defined according to the common use in engineering. Throughout this monograph the above relation is expressed as $\varepsilon = \varepsilon(u)$.

(c) The initial and boundary conditions :

For simplicity's sake we assume that $u_i = \sigma_i = 0$ at t = 0. In the dynamic problem, we further assume that $\dot{u}_i(0, x) = a_i(x)$, where $a_i(x)$ has suitable smoothness and compatibility conditions. Also in the quasi-static problem, we need the compatibility condition $b_i(0,x) = 0$.

Now let us consider the case where the deformation proceeds beyond the elastic limit. We assume that this limit is given by the condition of von Mises :

$$(3.1) \qquad f^2(\sigma) = \sigma_{11}^2 + \sigma_{22}^2 - \sigma_{11}\sigma_{22} + 3\sigma_{12}^2 = z_0^2$$

We also assume, as in the mass system, that the strain increments are the sum of the elastic and plastic increments :

$$(3.2) \qquad \dot{\varepsilon} = \dot{\varepsilon}^e + \dot{\varepsilon}^p,$$

and $\dot{\varepsilon}^e$ is connected with $\dot{\sigma}$ by the relation $\dot{\sigma} = D\dot{\varepsilon}^e$, and $\dot{\varepsilon}^p$ by

$$(3.3) \qquad \dot{\varepsilon}^p = \frac{1}{\eta}\,\partial f\,\partial f^*\dot{\sigma} \qquad (\,\partial f^*\dot{\sigma} \geq 0\,)$$

for a certain positive function η. Here ∂f is a vector defined by

$$\partial f^* = (\,\partial f/\partial\sigma_{11}\ \ \partial f/\partial\sigma_{22}\ \ \partial f/\partial\sigma_{12}),$$

which is parallel to the normal to the surface $f(\sigma) = z_0$ at the stress point σ , and D is the usual 3 × 3 constant matrix to represent a generilized Hooke's law.

(d) The hardening assumption :

The stress-strain relation in the plastic state is derived as follows.

(1) The isotropic hardening case . We assume that the hardening is expressed by a monotone function of the plastic work :

$$(3.4) \qquad f(\sigma) = F(w^p) \qquad w^p = \int_0^t \sigma^*\dot{\varepsilon}^p dt.$$

Introducing the parameter $\bar{\varepsilon}^p$ defined by

$$\dot{\bar{\varepsilon}}^p = \frac{\dot{w}^p}{f(\sigma)}$$

the function F is represented by a certain monotone function H as

$$(3.5) \qquad f(\sigma) = H(\bar{\varepsilon}^p) \qquad \bar{\varepsilon}^p = \int_0^t \dot{\bar{\varepsilon}}^p \, dt.$$

This is justified in the same way as before. Therefore once this H is given, then the η in (3.3) is obviously equal to $H'(\bar{\varepsilon}^p)$.

Now, since $\dot{\varepsilon}^e = \dot{\varepsilon} - \dot{\varepsilon}^p$ and $\dot{\sigma} = D\dot{\varepsilon}^e$, we have by (3.2)

$$(3.6) \qquad \dot{\sigma} = D\dot{\varepsilon} - D\dot{\varepsilon}^p = D\dot{\varepsilon} - \frac{1}{H'} D\partial f \partial f^* \dot{\sigma}.$$

Hence we have

$$\partial f^* \dot{\sigma} = \partial f^* D\dot{\varepsilon} - \frac{1}{H'} \partial f^* D \, \partial f \cdot \partial f^* \dot{\sigma},$$

so that $\partial f^* \dot{\sigma} = (1 + 1/H' \cdot \partial f^* D \partial f)^{-1} \partial f^* D\dot{\varepsilon}$. Substituting this into (3.6), we have

$$(3.7) \qquad \dot{\sigma} = \left(D - \frac{D\partial f \partial f^* D}{H' + \partial f^* D \partial f} \right) \dot{\varepsilon}.$$

The stress-strain relation in the case of isotropic hardening is therefore

$$(3.8) \quad \begin{cases} \dot{\sigma} = D\dot{\varepsilon}, \ \dot{\varepsilon}^p = 0 & \text{if } f(\sigma) < H(\bar{\varepsilon}^p) \text{ or } f(\sigma) = H(\bar{\varepsilon}^p) \text{ and } \partial f^* \dot{\sigma} < 0 \\ \dot{\sigma} = \left(D - \dfrac{D\partial f \partial f^* D}{H' + \partial f^* D \partial f} \right) \dot{\varepsilon}, \ \dot{\varepsilon}^p = \dfrac{1}{H'} \partial f \partial f^* \dot{\sigma} & \text{if } f(\sigma) = H(\bar{\varepsilon}^p) \\ & \text{and } \partial f^* \dot{\sigma} \geq 0, \end{cases}$$

where $\bar{\varepsilon}^p$ is defined by (3.5). The condition " or $f(\sigma) = H(\bar{\varepsilon}^p)$ and $\partial f^* \dot{\sigma} < 0$ " in (3.8) is not necessary for the dynamic problem. Also H' can be replaced by $g(\sigma) = H'(H^{-1}(f(\sigma)))$. We use this expression in what follows.

REMARK : The quantity $\dot{\bar{\varepsilon}}^p$ is called the " rate of equivalent plastic strain ", and in the general theory of plasticity it is connected with the strain increment $\dot{\varepsilon}^p$ by

$$\dot{\bar{\varepsilon}}^p = \sqrt{\frac{2}{3}}\,\|\dot{\varepsilon}^p\|_{R^6} .$$

This relation, however, does not hold for two-dimensional problems -- that is, for the three - dimensional expression of the stress and strain.

(2) The kinematic hardening case . We employ Ziegler's rule [20]. Let α be the parameter to represent the center of the yield surface with components α_{ij}:

$$\alpha = (\alpha_{11} \quad \alpha_{22} \quad \alpha_{12}).$$

The yield surface then takes the form

$$f(\sigma - \alpha) = z_0$$

The direction of $\dot{\alpha}$ is assumed to be parallel to the vector $\sigma - \alpha$, so that

(3.9) $$\dot{\alpha} = (\sigma - \alpha)\mu$$

for a certain nonnegative function μ . This function is determined automatically by the condition that the stress point lies on the yield surface during the plastic deformation :

$$\partial f^*(\dot{\sigma} - \dot{\alpha}) = 0 \qquad (\partial f = \frac{\partial}{\partial\sigma} f(\sigma - \alpha)).$$

In fact, taking into account the identity $\partial f^*(\sigma - \alpha) = f$, we have

(3.10) $$\dot{\alpha} = (\sigma - \alpha)\frac{\partial f^*\dot{\sigma}}{f}$$

If we assume that the plastic strain increments are given by

(3.11) $$\dot{\varepsilon}^p = \frac{1}{\eta}\,\partial f \partial f^* \dot{\sigma}$$

then the stress-strain relation is given, as in the isotropic case, by

$$\dot{\sigma} = \left(D - \frac{D\partial f \partial f^* D}{\eta + \partial f^* D \partial f} \right) \dot{\varepsilon} \; . \tag{3.12}$$

As η, one can assign H' determined just as in the isotropic case. This choice may be interpreted as follows.

The identity $f(\sigma - \alpha) = z_0$ in the plastic state is caused by the translation of the center of the yield surface. Therefore there must exist a substantial increase of the stress. Thus we assume that this part of the increasing is a monotone function of $\bar{\varepsilon}^p$, that is,

$$z_0 + \int_p \partial f^* \dot{\sigma} \, dt = H\left(\int_0^t \dot{\bar{\varepsilon}}^p dt \right), \tag{3.13}$$

where $\int_p$ denotes the integration in the plastic state, and

$$\dot{\bar{\varepsilon}}^p = \frac{\dot{w}^p}{f(\sigma - \alpha)} \qquad w^p = \int_0^t (\sigma - \alpha)^* \dot{\varepsilon}^p \, dt.$$

Then the function η must be equal to $H'(\bar{\varepsilon}^p)$ because we have in the plastic state

$$\partial f^* \dot{\sigma} = H' \dot{\bar{\varepsilon}}^p = H' \frac{(\sigma - \alpha)^*}{f(\sigma - \alpha)} \frac{1}{\eta} \partial f \cdot \partial f^* \dot{\sigma} = \frac{H'}{\eta} \partial f^* \dot{\sigma} \; .$$

The stress-strain relation in the kinematic hardening case is therefore written as

$$\left\{ \begin{array}{ll} \dot{\sigma} = D\dot{\varepsilon} \; , \quad \dot{\alpha} = 0 & \text{if } f(\sigma - \alpha) < z_0 \quad [\text{ or } f(\sigma - \alpha) = z_0 \text{ and } \partial f^* \dot{\sigma} < 0 \,] \\ \dot{\sigma} = \left(D - \dfrac{D\partial f \partial f^* D}{\eta + \partial f^* D \partial f} \right) \dot{\varepsilon} \; , \quad \dot{\alpha} = (\sigma - \alpha) \dfrac{\partial f^* \dot{\sigma}}{f(\sigma - \alpha)} & \text{if } f(\sigma - \alpha) = z_0 \\ & \text{and } \partial f^* \dot{\sigma} \geq 0. \end{array} \right. \tag{3.14}$$

The condition in the blanket is not necessary for the dynamic problem. We assume, as stated before, that η is constant for the kinematic case.

In this chapter we sketched briefly the formulation of some fundamental problems of plasticity. The following chapters are devoted to analysing them from some mathematical points of view.

CHAPTER 2

ELASTIC-PLASTIC VIBRATION OF A SPRING-MASS SYSTEM WITH ONE DEGREE OF FREEDOM

2.1. Function of bounded right derivative

The stress velocity $\dot{\sigma}$ or the strain velocity $\dot{\varepsilon}$ in the quasi-static problem is not necessarily continuous. However, as is proved later, the right derivative of these functions exist in a wide class of problems. Accodingly, we shall introduce a family of functions C_+^1 in which the stress and strain velocities are sought. Let (a,b) be a bounded, open interval.

DEFINITION. $C_+^1(a,b)$ is the set of functions which are continuous on (a,b) and have a uniformly bounded right derivative at every point of (a,b). For $\phi(x) \in C_+^1(a,b)$ we define

$$\phi'_+(x) = \lim_{\Delta x \to +0} \frac{\phi(x+\Delta x) - \phi(x)}{\Delta x} \qquad x \in (a,b)$$

$$|\phi|_1^+ = \sup_{x\in(a,b)} |\phi'_+(x)|.$$

THEOREM 2.1. For any $\phi(x) \in C_+^1(a,b)$ it holds that

$$(1.1) \qquad \left| \frac{\phi(\alpha) - \phi(\beta)}{\alpha - \beta} \right| \leq |\phi|_1^+ \qquad \text{for all } \alpha, \beta \in (a,b) \quad (\alpha < \beta).$$

Hence $\phi(x)$ is absolutely continuous.

PROOF. Assume that $\phi(\alpha) < \phi(\beta)$. Then it is easy to see that there are

$\gamma_1, \gamma_2 \in [\alpha,\beta]$ such that

$$\phi'_+(\gamma_1) \leq \frac{\phi(\beta) - \phi(\alpha)}{\beta - \alpha} \leq \phi'_+(\gamma_2). \tag{1.2}$$

Hence (1.1) holds. The situation is the same when $\phi(\beta) < \phi(\alpha)$. This completes the proof.

In this monograph the derivative means the right derivative when the ordinal one does not exist, and we assume that all functions belong to the class C^1_+ when their derivatives come into question, unless otherwise specified.

2.2. Equation of motion (kinematic hardening problem)

The initial value problem of the single mass system formulated in Sec.1 of Chapter 1 was to solve the system of equations

$$\rho\ddot{u} + \sigma = b \qquad \text{in } I \tag{2.1}$$

$$\begin{cases} \dot{\sigma} = k\dot{u}, \quad \dot{\alpha} = 0 & \text{if } |\sigma - \alpha| < z_0 \\ \dot{\sigma} = (1 - \xi)k\dot{u}, \quad \dot{\alpha} = \dot{\sigma} & \text{if } |\sigma - \alpha| = z_0 \text{ and } (\sigma - \alpha)\dot{\sigma} \geq 0 \end{cases} \tag{2.2}$$

under the initial conditions $u(0) = 0$, $\dot{u}(0) = a$, $\sigma(0) = \alpha(0) = 0$ and the initial yield condition $|\sigma| = z_0$. The first fundamental question is whether this problem can be well set up as an initial value problem of ordinary differential equations. To answer this question we reformulate this problem into a series of " elastic problems . " Such reformulation makes the problem very simple, and as a result we can find out some essential characters of the elastic-plastic deformation.

The stress-strain relation (2.2) can be interpreted as follows. Let us consider the two lines in (u,σ)-space (see Fig.1) :

(2.3) $$P^{\pm} : \sigma \mp z_0 = (1 - \xi)k(u \mp \frac{1}{k} z_0).$$

We call the union of these sets the plastic zone. On the other hand, the elastic zone is defined by

$$E = \{ (u,\sigma) \; ; \; (1 - \xi)ku - \xi z_0 < \sigma < (1 - \xi)ku + \xi z_0 \}$$

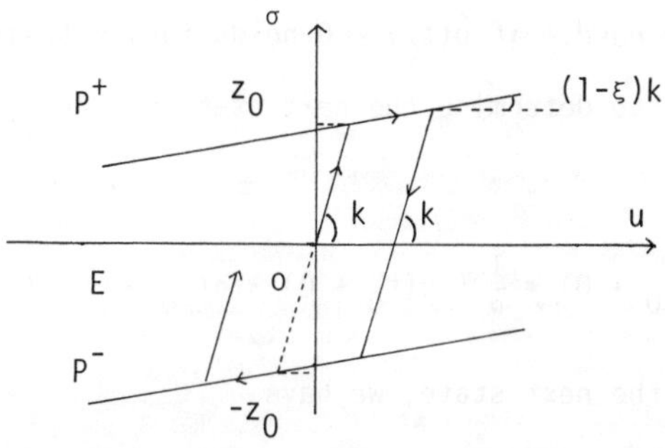

Fig. 1 The elastic and plastic zones

It is evident that the system is elastic if $(u, \sigma) \in E$ and plastic if $(u,\sigma) \in P^+ \cup P^-$. We say that the stress-strain relation (or the solution of (2.1)) is admissible if the following conditions are satisfied.

(1) $\sigma \in C^1_+(I)$ and $(u,\sigma) \in E \cup P^+ \cup P^-$.

(2) The relation $\dot{\sigma} = k\dot{u}$ (or $\dot{\sigma} = (1 - \xi)k\dot{u}$) holds in E (or on $P^{\pm}$).

(3) The point (u, σ) can move only to the right on P^+ and to the left on P^-.

Let us construct a solution of (2.1)$\sim$(2.2) under the above rule. We start from t = 0 with the given initial condition. In a certain neighborhood of t = 0, the solution is clearly small, and thus the elastic stress - strain relation must be used. Hence there exists a unique solution of the present problem until the time $t = t_0$ at which the condition $|\sigma| = z_0$ is satisfied. Set $\sigma_0 = \sigma(t_0)$ and $u_0 = u(t_0)$. We assume $u_0 > 0$, for example. If $\dot{u}(t_0) \neq 0$, then the system is clearly plastic after t_0 since the solution of (2.1) is increasing at t_0 for any choice of the next stress - strain relation. In this case, therefore, (2.1) is integrated under the relation $\dot{\sigma} = (1 - \xi)k\dot{u}$ - that is, $\sigma = (1-\xi)ku + \xi ku_0$ - and the solution ofour problem is continuated beyond $t = t_0$.

On the other hand, if $\dot{u}(t_0) = 0$ holds incidentally, we need the sign of $\ddot{u}(t_0 + 0)$ in order to determine the next state of the system. In this case, since

$$\ddot{u}(t_0 + 0) = \frac{1}{\rho} [\, b(t_0 + 0) - \sigma(t_0)] \qquad (\equiv C_2)$$

for any choice of the next state, we have

(2.4) if $C_2 > 0$ (or < 0) then σ is increasing (or decreasing) after t_0.

Hence, if $C_2 \neq 0$ then the stress-strain relation after t_0 is admissible with

" plastic (or elastic) if $C_2 > 0$ (or < 0), "

and the solution of our problem is continuated in this case too. Note that we can always get a solution of (2.1) under an arbitrary stress-strain relation. But it is another matter whether this solution is admissible, since the solution must satisfy the conditions (1)~(3) above.

Now if $C_2 = 0$, then we have to examine the higher derivative of u. Clearly

(2.5) $$\frac{d^3}{dt^3} u(t_0 + 0) = \frac{1}{\rho} [\dot{b}(t_0 + 0) - \dot{\sigma}(t_0 + 0)]$$

for any choice of the state after t_0. Now since $\dot{u}(t_0) = 0$, we have $\dot{\sigma}(t_0 + 0) = 0$ for any choice of the next state. Hence (2.5) implies that the value

$d^3 u(t_0+0)/dt^3$ is determined independently of the choice of the next state. Therefore, if we denote the right side of (2.5) by C_3, then (2.4) holds replacing C_2 by C_3 and so there exists an admissible σ - u relation. The solution is hence continuated across t_0 in this case too. Furthermore, if $C_3 = 0$ we can repeat this argument. In general we have

LEMMA 2.2. Let $u^{(m)}$ be the m-th order derivative of u. Assume that the following conditions are satisfied by the solution of the present problem.

(a) $u^{(i+2)}(t_0+0)$ $(i \leq m)$ are determined independently of the choice of the next state, and

(b) $u^{(i+1)}(t_0+0) = 0$ $(i \leq m)$.

Then we have

(A) the signs of $\sigma^{(i+2)}(t_0+0)$ $(i \leq m)$ are determined independently of the choice of the next state, and

(B) if $\sigma^{(m+2)}(t_0+0) = 0$, then the above (a) and (b) hold replacing m by m+1.

PROOF. Solve the equation (2.1) by assuming that the next state is elastic (or plastic). We then have

$$\sigma^{(i+2)} = ku^{(i+2)} \quad (\text{or} \; = (1-\xi)ku^{(i+2)}) \qquad (i \leq m) \tag{2.6}$$

at $t = t_0+0$. By the assumption (a), $u^{(i+2)}(t_0+0)$ $(i \leq m)$ is independent of the next state and so follows (A). To prove (B) we note that the following equality holds independently of the next state.

$$u^{(m+3)}(t_0+0) = \frac{1}{\eta}\,[\,b^{(m+1)}(t_0+0) - \sigma^{(m+1)}(t_0+0)\,] .$$

Since $\sigma^{(m+1)}(t_0+0) = 0$ by (b), the value $u^{(m+3)}(t_0+0)$ is independent of the next state. Hence (a) holds replacing m by m+1. $u^{(m+2)}(t_0+0) = 0$ follows from (2.6) and the assumption. This completes the proof.

We required the piecewise analiticity of the function b. Therefore there must exist m such that $b^{(m)}(t_0+0) \neq 0$, provided b is not constant for $t \geq t_0$. After all, this lemma implies that at $t = t_0+0$ the value of the derivative of the lowest order which is not zero is determined independently of the choice of the next state. Therefore we can determine the next state by examining the sign of this value.

Now assume that the order of this derivative is k_0 and $u^{(k_0)}(t_0+0) > 0$. The next state is then plastic, since we assumed $u_0 > 0$. Hence the equation (2.1) is solved with the relation $\sigma = (1-\xi)k(u-u_0) + z_0$. Suppose that at $t = t_1$, $\dot{u}(t) = 0$ holds at the first time after t_0. Here arises again the problem of how to determine the next state. The situation, however, is exactly the same as in the case of $t = t_0$. In fact, we can find the smallest k_1 such that $u^{(k_1)}(t_0+0) \neq 0$ if b is not constant. Thus the next state is

" plastic (or elastic) if $u^{(k_1)}(t_1+0) > 0$ (or < 0) . "

Now suppose that $u^{(k_1)}(t_1+0) < 0$, that is, that the unloading occurs at t_1. Set $\sigma_1 = \sigma(t_1)$, $u_1 = u(t_1)$. Then the equation (2.1) is solved with the relation

$$\sigma = k(u - u_1) + \sigma_1.$$

Therefore, if the displacement in the opposite direction is very large, the point (u, σ) can reach to the line P^- in (u,σ)-plane. The above discussion is completely valid in this case too, and thus an admissible σ - u relation is obtained for the next time interval. Since the boundedness of the solution is always assured, as is seen in the next section, the solution can be continuated all over the interval I.

Summarizing the results, we have

THEOREM 2.3 Suppose that $(u(t_0), \sigma(t_0)) \in P^+ \cup P^-$ and $\dot{u}(t_0) = 0$ for the solution u of (2.1) in $(0, t_0)$ with admissible σ - u relation. Then the solution of (2.1) is either constant for $t > t_0$ or there exists k_0 $(< \infty)$ such that

$$u^{(k)}(t_0 + 0) = 0 \quad (k < k_0) \quad \text{and} \quad u^{(k_0)}(t_0 + 0) \neq 0.$$

Hence the next state can be determined by the sign of this value.

2.3. Energy inequalities

Let $u_{(j)}$ be the displacement where the (j+1)-th change of the state occurs. If $u_{(m)}$ is the last such displacement we call that the system is in stage(m). (If, however, $(u,\sigma)\in P^+ \cup P^-$ at $t = t'$ but $(u,\sigma) \in E$ before and after t', then we think that the system keeps to be elastic).

The equations (2.1)~(2.2) can be represented by a single equation.

THEOREM 2.4. If the σ - u relation is admissible then the equation of motion (2.1) is represented as follows in stage(m).

$$\rho\ddot{u} + ku - \xi k \sum_{j=0}^{m} (-1)^j (u - u_{(j)}) = b. \tag{3.1}$$

PROOF. We use an induction on m. The theorem is correct for m=0. Assume that (3.1) holds until m = r (≥ 0).

If stage(r) is elastic then $\dot{\sigma}$ is given by $\dot{\sigma} = (1 - \xi)k\dot{u}$ in stage(r+1). Let t_{r+1} be the time from which stage(r+1) begins. Integrating $\dot{\sigma} = (1 - \xi)k\dot{u}$ in stage(r+1), we have

$$\sigma(t) = \sigma(t_{r+1}) + (1 - \xi)k(u - u_{(r+1)}). \tag{3.2}$$

On the other hand, by the assumptions of the induction, the following equality holds in stage(r).

$$\sigma(t) = ku - \xi k \sum_{j=0}^{r} (-1)^j (u - u_{(j)}).$$

Therefore, by the continuity of σ, it holds that

$$\sigma(t_{r+1}) = ku_{(r+1)} - \xi k \sum_{j=0}^{r} (-1)^j (u_{(r+1)} - u_{(j)})$$

$$= ku_{(r+1)} + \xi k \sum_{j=0}^{r} (-1)^j u_{(j)} \qquad \text{(r: odd)}.$$

Substituting this into (3.2), we have

$$\begin{aligned}\sigma(t) &= ku_{(r+1)} + \xi k \sum_{j=0}^{r} (-1)^j u_{(j)} + (1-\xi)k(u - u_{(r+1)}) \\ &= (1-\xi)ku + \xi k \sum_{j=0}^{r+1} (-1)^j u_{(j)} \\ &= ku - \xi k \sum_{j=0}^{r+1} (-1)^j (u - u_{(j)}).\end{aligned}$$

Therefore (3.1) holds for $m = r+1$. The situation is completely the same also in the case that stage(r) is plastic. The induction is hence complete.

By this expression we can easily derive a simple equality which represents the damping effect due to the plastic deformation. Let E_m be defined by

(3.3) $$E_m(t) = \frac{\rho}{2}(\dot{u})^2 + \frac{k}{2}u^2 - \frac{\xi k}{2}\sum_{j=0}^{m} (-1)^j (u - u_{(j)})^2 - \int_0^t b\dot{u}dt.$$

THEOREM 2.5. The following equality holds in stage(m).

(3.4) $$E_m(t) = \frac{\rho}{2}(\dot{u})^2(0) \qquad (\ = \text{initial energy}\ ).$$

PROOF. Let t_m be the time from which stage(m) begins. Multiplying both sides of (3.1) by $\dot{u}$ and integrating from t_m to t in stage(m), we have

$$\begin{aligned}E_m(t) &= \frac{\rho}{2}(\dot{u})^2(t_m) + \frac{k}{2}(u_{(m)})^2 - \frac{\xi k}{2}\sum_{j=0}^{m-1} (-1)^j (u_{(m)} - u_{(j)})^2 - \int_0^{t_m} b\dot{u}\,dt \\ &= E_{m-1}(t_m),\end{aligned}$$

which is the limit of E_{m-1} in the preceding stage. This implies that the quantity defined by (3.3) is constant through all stages. Hence (3.4) follows.

REMARK. We conjecture that the number of the state changes is finite in a finite time interval, provided the function b has good properties. However, this is not yet proved, and so we have to consider the case where $\{u_{(j)}\}$ has accumulation points. Even in such a case, however, the situation is the same. To see this, let $t_0 < t_1 < \dots$ be the time at which the state changes, and set $u_{(j)} = u(t_j)$. Let $\bar{t}$ be the first accumulation point of the above t_j' s. We assume that $\dot{u}$ is bounded in any finite time interval (this is proved later). Then, since the number of the hysteresis loops is finite in finite time interval, we can find a suitable unloading displacement $u_{(i)}$ such that all the subsequent unloading displacements $u_{(i+2k')}$ and the yield displacement $u_{(i+2k'+1)}$ are equal (see Fig. 2) :

$$u_{(i+2k')} = u_{(i+2k'+1)} \qquad \text{for all } k' \geq 0$$

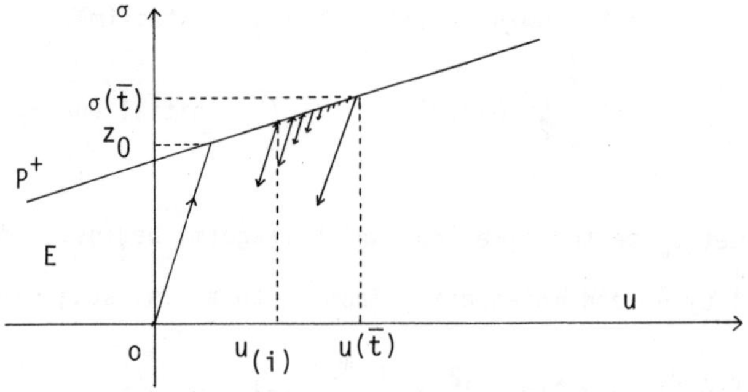

Fig. 2. A " pathological " case

Therefore we have at $t = \bar{t}$

$$\sigma(\bar{t}) = ku(\bar{t}) - \xi k \sum_{j=0}^{i-1} (-1)^j (u(\bar{t}) - u_{(j)}).$$

In other words, we can skip all the numbers after i to represent $\sigma(t)$ after $\bar{t}$.

In fact, if the first state after $\bar{t}$ is plastic, then we have for t near $\bar{t}$

$$\sigma(t) = ku(t) - \xi k \sum_{j=0}^{i-1} (-1)^j (u(t) - u_{(j)}) \qquad (t > \bar{t});$$

and if it is elastic, then similarly we have

$$\sigma(t) = ku(t) - \xi k \sum_{j=0}^{i} (-1)^j (u(t) - u_{(j)}) \qquad (t > \bar{t}),$$

where we redefine $u_{(i)} = u(\bar{t})$. Note that this simplification is possible also for the expression of $E_m(t)$ defined by (2.3). In what follows, we assume that $u_{(j)} \neq u_{(j+1)}$ for all $j \leq m-1$ whenever we consider the system after stage(m).

REMARK. The quantity

$$E^*(t) = -\frac{\xi k}{2} \sum_{j=1}^{m} (-1)^j (u - u_{(j)})^2$$

is regarded as a certain amount of energy used to cause plastic deformation.

To see this, consider first the quantity

$$e_{k'}(v) = -\frac{\xi k}{2} \sum_{4k'+1}^{4k'+4} (-1)^j (v - u_{(j)})^2 \qquad (k' \geq 0,\ 4k' + 4 \leq m)$$

for arbitrary real number v. It is easy to see that $e_{k'}(v)$ is a constant function of v. Also $e_{k'}(u_{4k'+4})$ is the area of the parallelogram (the hysteresis loop) formed by the points $P_j = (u_{(4k'+j)}, \sigma_{(4k'+j)})$ $(j = 1 \sim 4)$, where $\sigma_{(j)}$ is the value of σ when $u = u_{(j)}$ holds (see Fig. 3). Now we partition the terms of $E^*(t)$ every four terms from the beginning. If m is not divisable by 4, then the possible remainder in $E^*(t)$ is $\xi k/2$ times one of the following quantities, which are non-negative (see Fig. 4):

$$(u - u_{(m)})^2, \quad (u - u_{(m-1)})^2 - (u - u_{(m)})^2$$

$$(u - u_{(m-2)})^2 - (u - u_{(m-1)})^2 + (u - u_{(m)})^2.$$

Hence E*(t) is non-negative and it is clear that this quantity expresses the decreasing of the energy of the system which is caused by the appearance of the hysteresis loops.

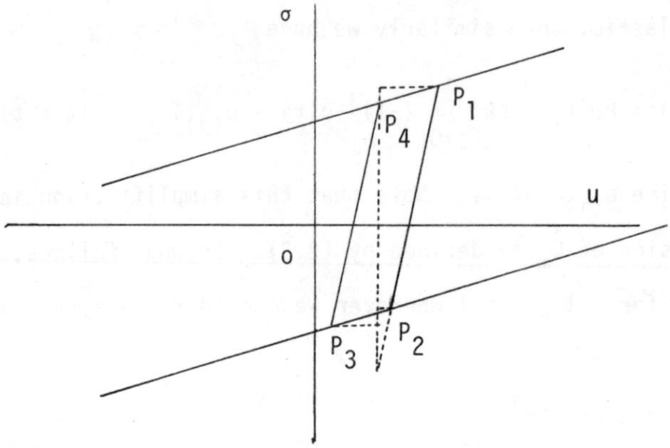

Fig. 3. $e_{k'}(v) = e_{k'}(u_{4k'+4})$ = area of the loop $P_1 \to P_4$.

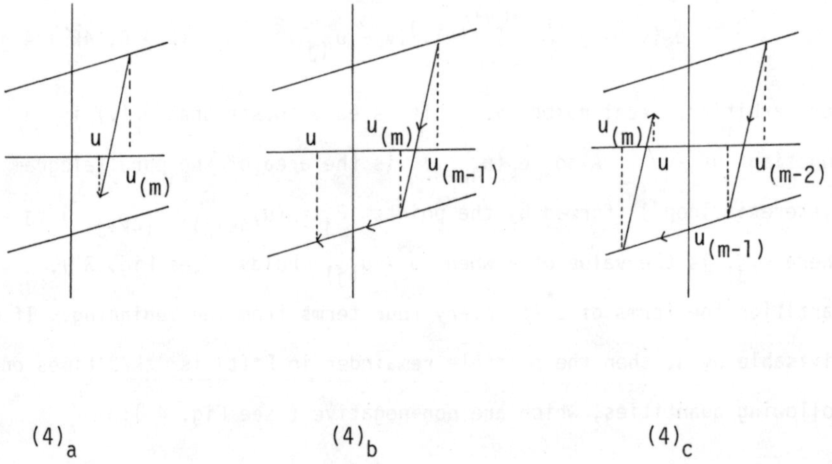

Fig. 4 The residual terms

THEOREM 2.6 (Energy inequalities). Assume that the solution is continuated up to $t = T_0$. Then for any $t \leq T_0$ it holds that

$$\frac{\rho}{2}(\dot{u})^2 + \frac{(1-\xi)k}{4} u^2 \leq [C_1 + \frac{1}{\sqrt{2\rho}} \int_0^{T_0} |b| \, dt]^2 \tag{3.5}$$

$$\frac{\rho}{2}(\ddot{u})^2 + \frac{(1-\xi)k}{2} \dot{u}^2 \leq [C_2 + \frac{1}{\sqrt{2\rho}} \int_0^{T_0} |\dot{b}| \, dt]^2, \tag{3.6}$$

where C_1 and C_2 are constants depending only on the given data.

PROOF. Since $E^*(t) \geq 0$, we have

$$\frac{k}{2} u^2 - \frac{\xi k}{2} \sum_{j=0}^{m} (-1)^j (u - u_{(j)})^2$$

$$= \frac{k}{2} u - \frac{\xi k}{2} (u - u_{(0)})^2 + E^*(t)$$

$$\geq \frac{(1-\xi)k}{4} u^2 - \frac{\xi(1+\xi)k}{2(1-\xi)} u_{(0)}^2$$

and hence we have by Theorem 2.5

$$\frac{\rho}{2} \dot{u}^2 + \frac{(1-\xi)k}{4} u^2 \leq \frac{\rho}{2} \dot{u}^2(0) + \frac{\xi(1+\xi)k}{2(1-\xi)} u_{(0)}^2 + \int_0^t b\dot{u} \, dt.$$

Let $E(t)$ be the square root of the left side and C be the sum of the first and second terms of the right side of this inequality, respectively. We then have

$$E(t)^2 \leq C^2 + \int_0^t g(s)E(s)ds \qquad (g(t) = \sqrt{\frac{2}{\rho}} |b|). \tag{3.7}$$

To solve this inequality, let $Z(t)$ be the right side of this inequality. Since $Z'(t) = g(t)E(t)$, we have $Z'(t) \leq g(t)Z(t)^{1/2}$. Hence we have $[Z(t)^{1/2}]' \leq g(t)/2$, from which (3.5) follows.

To prove (3.6), differentiate both sides of (3.1) to get

$$\rho \frac{d^3}{dt^3} u + k\dot{u} - \xi k \sum_{j=0}^{m} (-1)^j \dot{u} = \dot{b}.$$

Assuming that the stage(m) begins from $t = t_0$, multiply both sides of this identity by $\ddot{u}$ and integrate the resulting equation from t_0 to t in stage(m). If we set

$$\bar{E}_m^2(t) \equiv \frac{1}{2}\rho\ddot{u}^2 + \frac{k}{2}\left(1 - \xi\sum_{j=0}^{m}(-1)^j\right)\dot{u}^2,$$

the result is written as

$$\bar{E}_m^2(t) - \bar{E}_m^2(t_0) = \int_{t_0}^{t}\dot{b}\,\ddot{u}\,dt.$$

Let t' be a point sufficiently near t_0 ($t' < t_0 < t$) so that there is no change of state between t' and t_0. Then we have

$$\bar{E}_{m-1}^2(t_0) - \bar{E}_{m-1}^2(t') = \int_{t'}^{t_0}\dot{b}\,\ddot{u}\,dt.$$

At $t = t_0$ the apparent difference of the energy is

$$\bar{E}_m^2(t_0) - \bar{E}_{m-1}^2(t_0) = \frac{-\xi k}{2}\,\dot{u}^2(t_0)\left[\sum_{j=0}^{m}(-1)^j - \sum_{j=0}^{m-1}(-1)^j\right]$$

$$= \frac{-\xi k}{2}\,(-1)^m\dot{u}^2(t_0) \le 0,$$

since $\dot{u}(t_0) = 0$ if m is odd. Therefore we have

$$\bar{E}_m^2(t) - \bar{E}_{m-1}^2(t') \le \int_{t'}^{t}\dot{b}\,\ddot{u}\,dt,$$

and, summing on m, we have

$$\bar{E}_m^2(t) \le \frac{\rho}{2}\ddot{u}^2(0) + \int_0^t\dot{b}\,\ddot{u}\,dt, \tag{3.8}$$

from which (3.6) follows.

2.4. Behaviour of the solution as $t \to \infty$

Since the boundedness of u and $\dot{u}$ is always assured by the energy estimate (3.5), the solution of (2.1) exists in any finite interval $I = (0,T)$ with an admissible stress-strain relation. This is obvious if the number of the state change is finite in finite time interval, but the existence is assured even if it is infinite. In fact, suppose that $t_1 < t_2 \ldots \to \bar{t}$, where t_i is the time at which the state changes (see Fig. 2). All results of the previous sections are valid at least for $t < \bar{t}$, and the estimate (3.5) is valid at $t = \bar{t}$ by the continuity of $\dot{u}$. Since the number of the hysteresis loops in $[t_1, \bar{t}]$ is finite, and $\ddot{u}$ is also finite by (3.6), it must hold that $\dot{u}(\bar{t}) = 0$. Hence we can apply Theorem 2.3 to continuate the solution beyond $\bar{t}$. It is clear that there is no bound beyond which this continuation is impossible.

To see the behaviour of the solution as $t \to \infty$, we prepare

LEMMA 2.7. Let C_1 and C_2 be nonnegative constants and suppose

$$E^2(t) \le C_1 + C_2 E(t) + \int_0^t f(s)E(s)ds, \tag{4.1}$$

where $E(t) \ge 0$, $f(t) \ge 0$, $E(t) \in C[0,\infty)$, $f(t) \in L^1(0,\infty)$. Then we have

$$E(t) \le C_1^{\frac{1}{2}} + C_2 + \int_0^t f(s)ds. \tag{4.2}$$

PROOF. We set

$$F(t) = | E(t) - \frac{C_2}{2} | .$$

Then by (4.1) we have

$$F^2(t) \le C_1 + \frac{C_2^2}{4} + \frac{C_2}{2} \int_0^t f(s)ds + \int_0^t f(s)F(s)ds.$$

Hence, by solving this inequality we have (4.2).

THEOREM 2.8. If one of the following conditions is satisfied, then the elastic-plastic vibration converges to an elastic vibration as $t \to \infty$.

(a) $\int_0^\infty |b(t)|dt < \infty$

(b) $\sup_I |b(t)| < \infty$ and $\int_0^\infty |\dot{b}(t)|\, dt < \infty$.

PROOF. If the number of the hysteresis loops is finite as $t \to \infty$, then the theorem is evident. Even if it is infinite, the width of the loops tends to 0 as $t \to \infty$. To see this, we first note that (as shown in the proof of Theorem 2.6), there is a constant C such that

$$\frac{\rho}{2}\dot{u}^2 + \frac{1}{4}(1 - \xi)ku^2 + E^*(t) \leq C + \int_0^t |b\dot{u}|\, ds.$$

Therefore if (a) is satisfied, we have by Lemma 2.7 $E^*(t) \leq$ const. as $t \to \infty$, so that the width of the loops tends to 0. On the other hand, if $b(t)$ is bounded, since

$$\int_0^t b\dot{u}\, ds = -\int_0^t \dot{b}u\, ds + [bu]_0^t,$$

there are two positive constants C_1 and C_2 such that

$$\frac{\rho}{2}\dot{u}^2 + \frac{1}{4}(1 - \xi)ku^2 + E^*(t) \leq C_1 + C_2|u| + \int_0^t |\dot{b}u|\, ds,$$

and hence we have again $E^*(t) \leq$ const. by Lemma 2.7.

2.5 A weak form of the stress-strain relation

The initial value problem (2.1)~(2.2) has another representation introduced originally by Duvaut-Lions in [4] and extended later by Johnson [9]. Define

$$K_\alpha = \{ \tau \in C^1_+(I) \ ; \ |\tau - \alpha| \leq z_0 \ \text{ for any } t \in I \}.$$

THEOREM 2.9. The initial value problem (2.1)~(2.2) is equivalent to the following problem : Seek $(u,\sigma,\alpha) \in C^1_+(I)$ which for all $t \in I$ satisfies the the following :

(5.1) $$\rho\ddot{u} + \sigma = b$$

(5.2) $$(\dot{\sigma} - k\dot{u}, \tau - \sigma) \geq 0 \qquad \text{for all } \tau \in K_\alpha,$$

(5.3) $$\dot{\alpha} = (1 - \frac{1}{\xi})(\dot{\sigma} - k\dot{u})$$

and $\sigma \in K_\alpha$, $u(0) = \sigma(0) = \alpha(0) = 0$, $\dot{u}(0) = a$.

REMARK. Here (x,y) denotes the product of the scaler functions x and y. This is only for the consistency of the expression. We later generalize the results of this theorem for more complicated cases.

PROOF OF THE THEOREM. Let α be the parameter to represent the center of the yield surface (in this case two points). We first show that the solution (u, σ) of (2.1)~(2.2) satisfies (5.2)~(5.3). If the system is elastic, then $\dot{\sigma} = k\dot{u}$ and $\dot{\alpha} = 0$, so that (5.2)~(5.3) holds well. If the system is plastic, $\dot{\sigma} = (1 - \xi)k\dot{u}$ and so $(\dot{\sigma} - k\dot{u}, \tau - \sigma) = -\xi k(\dot{u}, \tau - \sigma)$. In this case,

$$\text{if } \dot{u} > 0, \text{ then } \sigma = \alpha + z_0,$$

$$\text{if } \dot{u} < 0, \text{ then } \sigma = \alpha - z_0.$$

Therefore ,in any case we have

$$-\xi k(\dot{u}, \tau - \sigma) \geq 0 \qquad \text{for any } \tau \in K_\alpha,$$

which proves (5.2). Also $\dot{\alpha} = \dot{\sigma} = (1 - \xi)k\dot{u} = (1 - \frac{1}{\xi})(\dot{\sigma} - k\dot{u})$. Therefore (5.3) holds too. The proof is complete if the uniqueness of the solution is shown. Substitute (5.3) into (5.2) to get

$$(\dot{\alpha}, \tau - \sigma) \leq 0 \qquad \text{for any } \tau \in K_\alpha .$$

Let θ be an arbitrary $C^1_+(I)$-function satisfying $|\theta| \leq 1$. Then, since $\tau \equiv \alpha + z_0\theta \in K_\alpha$, we have

(5.4) $$(\dot{\alpha}, \alpha + z_0\theta - \sigma) \leq 0 \qquad \text{for any such } \theta .$$

Suppose now that there is another solution $(u_*, \sigma_*, \alpha_*)$. Obviously we have

(5.5) $$(\dot{\alpha}_*, \alpha_* + z_0\theta - \sigma_*) \leq 0 \qquad \text{for any such } \theta .$$

Set $\theta = (\sigma_* - \alpha_*)/z_0$ in (5.4) and $\theta = (\sigma - \alpha)/z_0$ in (5.5) and add both inequalities to get

$$(\dot{\alpha} - \dot{\alpha}_*, \alpha - \alpha_* - (\sigma - \sigma_*)) \leq 0.$$

By using (5.1) and (5.3) we have

$$(\dot{\alpha} - \dot{\alpha}_*, \sigma - \sigma_*) = (1 - \frac{1}{\xi})(\frac{1}{2}|\sigma - \sigma_*|^2_t + \frac{\rho k}{2}|\dot{u} - \dot{u}_*|^2_t).$$

Hence we have

$$|\alpha - \alpha_*|^2 - (1 - \frac{1}{\xi})(|\sigma - \sigma_*|^2 + \rho k|\dot{u} - \dot{u}_*|^2) \leq 0$$

which implies the uniqueness of the solution.

REMARK. Let σ and α be functions in $C^1_+(I)$, and define

$$E = \{ t \in I ; |\sigma - \alpha| < z_0\}$$

$$P = \{ t \in I ; |\sigma - \alpha| = z_0\}.$$

The inequalities (5.2)~(5.3) then imply the following :

$$(5.6)\qquad \begin{cases} \dot\sigma = k\dot u & \text{on } E \\ \dot\sigma = (1-\xi)k\dot u \quad \text{and} \quad (\sigma-\alpha)\dot\sigma \geq 0 & \text{a.e. } P. \end{cases}$$

The first equality is obvious, since if $t \in E$ then $\tau(t) - \sigma(t)$ can take both positive and negative values. On the other hand, if $t \in P$ then the sign of $\dot\sigma - k\dot u$ is either zero or the same as that of $\alpha - \sigma$; that is, there is a nonnegative function λ such that

$$(5.7)\qquad \dot\sigma - k\dot u = \lambda \frac{\alpha-\sigma}{|\alpha-\sigma|}\ .$$

Since σ and α are absolutely continuous, we have for almost all $t \in P$

$$0 \geq \lim_{\Delta t \to +0} \{ [\sigma(t+\Delta t) - \alpha(t+\Delta t)]^2 - [\sigma(t) - \alpha(t)]^2 \} / \Delta t$$
$$= 2[\sigma(t) - \alpha(t)][\dot\sigma(t) - \dot\alpha(t)]$$

$$0 \leq \lim_{\Delta t \to +0} \{ [\sigma(t) - \alpha(t)]^2 - [\sigma(t-\Delta t) - \alpha(t-\Delta t)]^2 \} / \Delta t$$
$$= 2[\sigma(t) - \alpha(t)][\dot\sigma(t) - \dot\alpha(t)],$$

so that

$$(5.8)\qquad \dot\sigma(t) - \dot\alpha(t) = 0 \qquad \text{a.e. } P.$$

Now we have, by (5.3) and (5.7),

$$\dot\alpha = \lambda \frac{\xi-1}{\xi} \frac{\alpha-\sigma}{|\alpha-\sigma|}\ .$$

Substituting this into (5.8), we determine λ as follows :

$$\lambda = \frac{\xi}{\xi-1} \frac{\alpha-\sigma}{|\alpha-\sigma|}\dot\sigma \quad (\geq 0).$$

The second equality of (5.6) is obtained by substituting this into (5.7).

REMARK. Elastic-perfectly-plastic problem

The case that $\xi = 1$ is called the perfectly plastic problem. The stress-strain relation in the plastic state is given by $\dot{\sigma} = 0$ in this case. The main results of the previous sections are also valid for this problem. Assume that we start from $t = 0$, and at $t = t_0$ we have $|\sigma| = z_0$ for the first time. If $\dot{u}(t_0) = \ddot{u}(t_0) = 0$, then the sign of $d^3u(t_0 + 0)/dt^3$ is determined by $\dot{b}(t_0+0)$, since

$$\dot{\sigma}(t_0 + 0) = \begin{cases} k\dot{u}(t_0 + 0) & \text{if } t > t_0 \text{ is elastic} \\ 0 & \text{if } t > t_0 \text{ is plastic.} \end{cases}$$

Moreover, if $d^3u(t_0 + 0)/dt^3$ vanishes, then we repeat the same argument. Hence Theorem 2.3 holds in this case too. Theorems 2.4～2.6 are valid as they are. If $b = 0$, the "shakedown" occurs after the first yielding as is easily seen from the identity

$$\frac{\rho}{2}\dot{u}^2 + \int_0^t \sigma\dot{u}\,dt = \frac{\rho}{2}\dot{u}^2(0).$$

The spring-mass system repeats elastic vibration after the first yielding (see Fig. 5). A similar phenomenon occurs even where the hardening exists, although the number of yieldings depends on the initial energy (see Fig. 6). In fact, numerical experiences suggest that there must exist a definite rule between the number of yieldings and the initial energy, if $b = 0$.(*)

On the shakedown, see Kachanov [11], for example.

(*) : Recently this rule was found out by M. Araki. See [Proceedings of the joint meeting on applied mathematics , Dec. 1983, Kyoto].

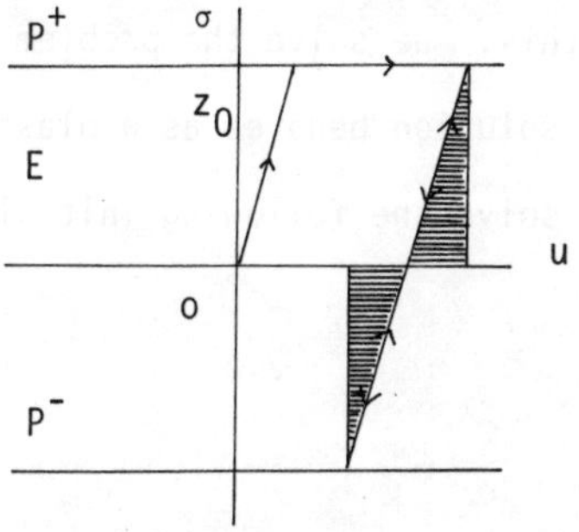

Fig. 5 Perfectly plastic case

(b = 0)

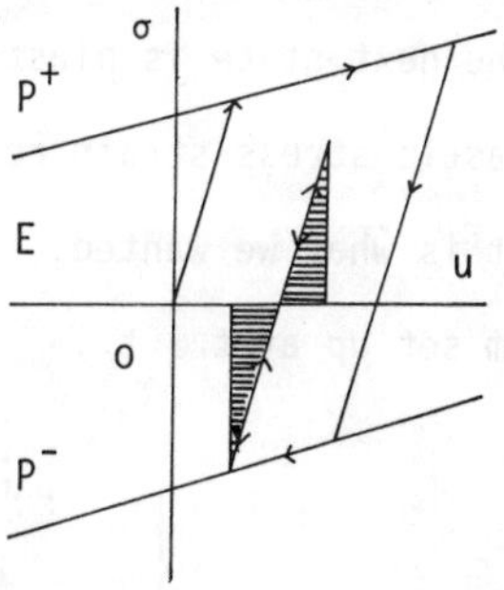

Fig.6 Shakedown

2.6. Isotropic hardening problem

The initial value problem of the single mass system with the isotropic hardening rule is to find (u,σ, u^p) which satisfies

(6.1) $\rho \ddot{u} + \sigma = b \quad \text{in } I$

(6.2) $\dot{\sigma} = k\dot{u}, \quad \dot{u}^p = 0 \quad \text{if } |\sigma| < H(\bar{u}^p)$

(6.3) $\dot{\sigma} = (k - \dfrac{k^2}{g(\sigma)+k})\dot{u}, \quad \dot{u}^p = \dfrac{1}{g(\sigma)}\dot{\sigma} \quad \text{if } |\sigma| = H(\bar{u}^p) \text{ and } \sigma\dot{\sigma} \geq 0,$

where $g(\sigma) = H'(H^{-1}(|\sigma|))$, $H(0) = z_0$ and

(6.4) $\bar{u}^p = \int_0^t |\dot{u}^p|\, dt, \quad u(0) = \sigma(0) = 0, \ \dot{u}(0) = a, \ u^p(0) = 0.$

In this case, the stress σ is not given explicitly as a function of u, and so it is impossible to express the equation of motion by using u. However we can prove the existence of a classical solution in a way similar to before.

Obviously the elastic rule is applied at $t = 0$. Assume that at $t = t_0$ $|\sigma| = H(0) = z_0$ hold. We want to determine the next state. If $\dot{u}(t_0) \neq 0$ then the next state is plastic. To verify this, we solve the problem under the plastic stress-strain relation. If the solution behaves as a plastic one, then it is what we wanted. Now, we have to solve the following initial value problem set up at $t = t_0$.

$$(6.5) \qquad \begin{cases} \rho\ddot{u} + \sigma = b \\ \dot{\sigma} = \left(k - \dfrac{k^2}{g(\sigma)+k}\right)\dot{u}, \end{cases}$$

where $u(t_0)$, $\dot{u}(t_0)$ and $\sigma(t_0)$ are given so as to keep the continuity of the solution. This system can be written as

$$(6.6) \qquad \frac{d}{dt}\begin{pmatrix} u \\ \dot{u} \\ \sigma \end{pmatrix} = X(t,\dot{u},\sigma)$$

for a suitable vector function X. This equation has a unique solution for the given initial condition, and $\sigma\dot{\sigma} > 0$ is satisfied at t_0 . To see this, we note that since $|\sigma(t_0)| = z_0$, $g(\sigma)$ is analytic with respect to t in a neighborhood of t_0 . Therefore, each component of X is analytic with respect to its arguments and, by the general theory of ordinally differential equations, (6.5) has a unique analytic solution in a neighborhood of t_0. Also the sign of $\dot{\sigma}(t_0 - 0)$ and of $\dot{\sigma}(t_0 + 0)$ are the same as that of $\dot{u}(t_0)$. Hence $\dot{u}(t_0) \neq 0$ implies $\sigma(t_0)\,\dot{\sigma}(t_0 + 0) > 0$, and $\sigma\dot{\sigma} > 0$ holds in a neighborhood of t_0. At the same time, as pointed out on page 8, the solution of (6.6) and hence of (6.5) satisfies

$$|\sigma| = H(\bar{u}^p) \qquad \text{where } \bar{u}^p(t) = \int_0^t |\dot{u}^p|dt \text{ and } \dot{u}^p = \frac{\dot{\sigma}}{g(\sigma)} .$$

Therefore the solution of (6.5) is what we wanted as the solution of (6.1)~

(6.3). This solution remains admissible as long as $\sigma \dot{\sigma} \geq 0$ is satisfied.

Now the question is where $\dot{u}(t_0) = 0$. If, however, $\ddot{u}(t_0) \neq 0$, then the next state can be determined by the following rule :

(*) If $\sigma(t_0) > 0$; the next state is plastic (or elastic) if $\ddot{u}(t_0) > 0$ (or < 0).

(*) If $\sigma(t_0) < 0$; the next state is elastic (or plastic) if $\ddot{u}(t_0) > 0$ (or < 0).

Here is why. We solve the equation (6.1) assuming that the next state is elastic or plastic. Since $\dot{u}(t_0) = 0$, the solution (u,σ) must satisfy the relation

$$\ddot{\sigma}(t_0 + 0) = k\ddot{u}(t_0 + 0) \qquad \text{if elastic,}$$

$$\ddot{\sigma}(t_0 + 0) = \left(k - \frac{k^2}{g(\sigma)+k}\right)\ddot{u}(t_0 + 0) \qquad \text{if plastic.}$$

Therefore, the sign of $\ddot{\sigma}(t_0 + 0)$ is determined independently of the choice of the next state, since $\ddot{u}(t)$ is continuous in either case. Now assume that $\sigma(t_0) > 0$. We then have

$$\ddot{\sigma}(t_0 + 0) > 0 \text{ if } \ddot{u}(t_0) > 0,$$

and hence $\sigma(t)$ is increasing after t_0 ; that is, $\sigma \dot{\sigma} > 0$. Hence the next state must be plastic. Similarly,

$$\ddot{\sigma}(t_0 + 0) < 0 \text{ if } \ddot{u}(t_0) < 0,$$

so that $\sigma(t)$ is decreasing after t_0. Hence the next state must be elastic. The rule when $\sigma(t_0) < 0$ is verified in the same way. Hence the next state is determined if $\ddot{u}(t_0) \neq 0$ and the solution is continuated beyond $t = t_0$. If $\ddot{u}(t_0) = 0$, or the higher derivatives vanish at t_0, we have to find the lowest derivative which does not vanish at t_0. The situation is the same as in the kinematic hardening case. In fact, we have

LEMMA 2.10. Assume that the solution is continuated up to $t = t_0$ and that the following conditions are satisfied.

(a) $u^{(i+2)}(t_0 + 0)$ $(i \leq m)$ is determined independently of the next state.

(b) $u^{(i+1)}(t_0 + 0) = 0$ $(i \leq m)$.

Then we have

(A) The sign of $\sigma^{(i+2)}(t_0 + 0)$ $(i \leq m)$ is determined independently of the next state.

(B) If $\sigma^{(m+2)}(t_0 + 0) = 0$, then (a) and (b) hold when m is replaced by m+1.

PROOF. We solve the problem assuming that the next state is elastic (or plastic). Then at $t = t_0 + 0$ the following equality must hold :

$$\sigma^{(m+2)} = ku^{(m+2)} \qquad (\text{or} \ = (k - \frac{k}{g(\sigma)+k})u^{(m+2)}).$$

By the assumption (a), $u^{(m+2)}(t_0 + 0)$ is independent of the the next state, so that (A) follows. To prove (B) we consider the equality

$$u^{(m+3)}(t_0 + 0) = \frac{1}{\rho} [b^{(m+1)}(t_0 + 0) - \sigma^{(m+1)}(t_0 + 0)]$$

which holds for any choice of the next state. By (b), $\sigma^{(m+1)}(t_0 + 0) = 0$, and so $u^{(m+3)}(t_0 + 0)$ depends only on b. Also, $\sigma^{(m+2)}(t_0 + 0) = 0$ implies that $u^{(m+2)}(t_0 + 0) = 0$. This proves (B).

Thus an admissible stress-strain relation is determined in the isotropic case . If the next state is elastic, then the stress-strain relation is as it is. If the next state is plastic, then this state continues as long as $\sigma\dot{\sigma} \geq 0$ is satisfied. When $\sigma\dot{\sigma} = 0$ is satisfied for a certain $t_1 > t_0$, the same problem occurs. In such a case, however, we can apply the above procedure to determine the next state.

ENERGY INEQUALITIES (isotropic hardening problem)

It suffices to get the estimate of the solution after a finite number of state changes, since we can apply the same argument as in the kinematic hardening case if it is infinite.

THEOREM 2.11. The following equality holds for any $t \in I$.

$$(6.7) \qquad \frac{1}{2}\left[\rho\dot{u}^2 + \frac{1}{k}\sigma^2\right]_0^t + \int_0^t \sigma\dot{u}^p\,dt = \int_0^t b\dot{u}\,dt.$$

PROOF. In the plastic state, we have

$$\dot{u} = \frac{g(\sigma)+k}{kg(\sigma)}\,\dot{\sigma},$$

so for any $t \in I$ it holds that

$$(6.8) \qquad \dot{u} - \frac{1}{k}\dot{\sigma} - \dot{u}^p = 0.$$

Multiply both sides of (6.8) by σ and integrate on t. Since $\rho\ddot{u}\dot{u} + \sigma\dot{u} = b\dot{u}$, we have the desired relation.

For the higher derivatives, we have

THEOREM 2.12. The following estimate holds for any $t \in I$:

(6.9) $$\frac{1}{2}\left[\rho\ddot{u}^2 + \frac{1}{k}\dot{\sigma}^2 + g(\sigma)(\dot{u}^p)^2\right]_0^t \leq \int_0^t \dot{b}\ddot{u}\,dt.$$

PROOF. Assume that the system is plastic on the interval $I_1 = (t_0, t_1)$. By (6.8) we have on I_1

(6.10) $$\ddot{u} - \frac{1}{k}\ddot{\sigma} - \left(\frac{d}{dt}\frac{1}{g(\sigma)}\right)\dot{\sigma} - \frac{1}{g(\sigma)}\ddot{\sigma} = 0.$$

Multiplying both sides by $\dot{\sigma}$, we have

(6.11) $$\left(\ddot{u} - \frac{1}{k}\ddot{\sigma}\right)\dot{\sigma} - \frac{1}{2}\frac{d}{dt}\left(\frac{1}{g(\sigma)}\dot{\sigma}^2\right) - \frac{1}{2}\left(\frac{d}{dt}\frac{1}{g(\sigma)}\right)\dot{\sigma}^2 = 0.$$

Let us introduce the quantity

$$E(t) = \frac{1}{2}\left(\rho\ddot{u}^2 + \frac{1}{k}\dot{\sigma}^2 + g(\sigma)(\dot{u}^p)^2\right)(t).$$

By the assumption on H we have

$$\frac{d}{dt}\left(\frac{1}{g(\sigma)}\right) = \frac{d}{dt}\left(\frac{1}{H'(\bar{u}^p)}\right) = \frac{-H''}{(H')^2}\,|\dot{u}^p| \geq 0,$$

and by (6.1)

$$\int_{t_0}^t \left(\frac{1}{2}\rho\frac{d}{dt}(\ddot{u})^2 + \dot{\sigma}\ddot{u}\right)dt = \int_{t_0}^t \dot{b}\ddot{u}\,dt \qquad t \in I_1.$$

Hence integrating (6.11) from t_0 to $t \in I_1$ we have

$$E(t) \leq E(t_0 + 0) + \int_{t_0}^t \dot{b}\ddot{u}\,dt.$$

Suppose now that the system is elastic in $I_0 = (t', t_0)$. Then, since it holds that at $t = t_0 + 0$

$$\frac{1}{k}(\dot{\sigma})^2 = \dot{\sigma}\left(\dot{u} - \frac{1}{g(\sigma)}\dot{\sigma}\right) = \dot{\sigma}\dot{u} - g(\sigma)(\dot{u}^p)^2,$$

we have

$$E(t_0+0) = \frac{1}{2}\rho\ddot{u}^2(t_0+0) + \frac{1}{2}\dot{\sigma}(t_0+0)\dot{u}(t_0+0)$$

$$= \frac{1}{2}\rho\ddot{u}^2(t_0+0) + \frac{1}{2}\dot{u}(t_0+0)(k - \frac{k^2}{g(\sigma)+k})\dot{u}(t_0+0)$$

$$\leq \frac{1}{2}\rho\ddot{u}^2(t_0+0) + \frac{k}{2}\dot{u}^2(t_0+0)$$

$$= \frac{1}{2}(\rho\ddot{u}^2(t_0-0) + \frac{1}{k}\dot{\sigma}^2(t_0-0))$$

$$= E(t_0-0),$$

so that the following estimate holds for any $t \in I_1$.

$$E(t) \leq E(t_0-0) + \int_{t_0}^{t} b\dddot{u}\,dt. \tag{6.12}$$

On the other hand, assume that the system is elastic on I_1. Then for any $t \in I_1$ we have

$$\frac{1}{2}(\rho\ddot{u}^2 + \frac{1}{k}\dot{\sigma}^2)(t) = \frac{1}{2}(\rho\ddot{u}^2 + \frac{1}{k}\dot{\sigma}^2)(t_0+0) + \int_{t_0}^{t} b\dddot{u}\,dt.$$

If the system is plastic on the interval $I_0 = (t', t_0)$, then $\dot{\sigma}(t_{0\pm}0) = 0$. Hence the inequality (6.12) holds in this case too. This proves the estimate of the theorem.

A WEAK FORM OF THE STRESS-STRAIN RELATION (isotropic hardening problem)

Since the function $H = H(\xi)$ ($\xi \in (-\delta, \infty)$) is monotonically increasing, we can define the one to one transformation $\xi \to \zeta$ by

$$\zeta = \zeta(\xi) = \int_0^{\xi} \sqrt{H'(\lambda)}\,d\lambda.$$

We introduce a new function $G(\zeta)$ of this new variable ζ as follows :

$$G'(\zeta) = \sqrt{H'(\xi)} \qquad G(0) = H(0). \tag{6.13}$$

Then we have $G(\zeta) = H(\xi)$, since for $\bar{\zeta} = \zeta(\bar{\xi})$ it holds that

$$G(\bar{\zeta}) = \int_0^{\bar{\zeta}} G'(\zeta)d\zeta + G(0) = \int_0^{\bar{\xi}} H'(\xi)d\xi + H(0) = H(\bar{\xi}).$$

The function G has the properties

$$G(0) > 0, \quad G'(0) \geq \sqrt{\delta_0} > 0, \quad G'' < 0,$$

on the new variable ζ, which were the properties of H. We now introduce a new parameter $\hat{u}^p$:

$$\hat{u}^p = \int_0^{\bar{u}^p} \sqrt{H'(\lambda)}\, d\lambda.$$

Then $G(\hat{u}^p) = H(\bar{u}^p)$, and G succeeds the properties of H . We define

$$B_H = \{(\tau,\zeta) \in R \times (-\delta\ ,\infty)\ ;\ |\tau| \leq H(\zeta)\},$$

$$B_G = \{(\tau,\zeta) \in R \times (-\delta^*,\infty)\ ;\ |\tau| \leq G(\zeta)\},$$

$$K_H = \{(\tau,\zeta) \in C_+^1(I) \times C_+^1(I)\ ;\ (\tau,\zeta) \in B_H \quad \text{for all } t \in I\},$$

$$K_G = \{(\tau,\zeta) \in C_+^1(I) \times C_+^1(I)\ ;\ (\tau,\zeta) \in B_G \quad \text{for all } t \in I\},$$

where $\delta^* = \int_{-\delta}^{0} \sqrt{H'(\lambda)}\, d\lambda$.

THEOREM 2.13. The initial value problem (6.1)~(6.3) is equivalent to the following problem : Seek $(u,\sigma,\hat{u}^p) \in C_+^1(I)$ which satisfies for all $t \in I$

(6.14) $$\rho\ddot{u} + \sigma = b,$$

(6.15) $$(\dot{u} - \frac{1}{k}\dot{\sigma}, \tau - \sigma) - (\dot{\hat{u}}^p, \zeta - \hat{u}^p) \leq 0 \qquad \text{for all } (\tau,\zeta) \in K_G,$$

and $(\sigma,\hat{u}^p) \in K_G$, $u(0) = \sigma(0) = \hat{u}^p(0) = 0$, $\dot{u}(0) = a$.

PROOF. We show that the solution $(u,\sigma,\hat{u}^p)$ of (6.1)~(6.3) satisfies these relations and that the solution of this problem is unique.

First, if the system is elastic, then $(\sigma, \hat{u}^p) \in K_G$ and it holds that

$$\dot{u} - \frac{1}{k}\dot{\sigma} = 0, \qquad \dot{\hat{u}}^p = \sqrt{H'(\bar{u}^p)}\,\dot{\bar{u}}^p = 0.$$

Hence (6.15) is satisfied. If it is plastic, then $(\sigma,\hat{u}^p)$ lies on the boundary of B_G, and the vector

$$\begin{pmatrix} \dot{u} - \frac{1}{k}\dot{\sigma} \\ -\dot{\hat{u}}^p \end{pmatrix} = \begin{pmatrix} \dot{u}^p \\ -\sqrt{H'(\bar{u}^p)}\,\dot{\bar{u}}^p \end{pmatrix} = \begin{pmatrix} \frac{\sigma}{|\sigma|} \\ -G'(\hat{u}^p) \end{pmatrix} \dot{\bar{u}}^p$$

is parallel to the outward normal to the closed convex set B_G at the boundary point $(\sigma,\hat{u}^p)$ (see Fig. 7). Hence (6.15) holds. The uniqueness of the solution is readily proved.

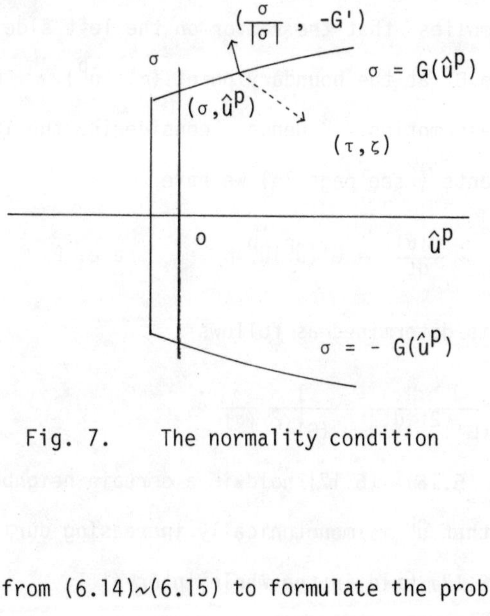

Fig. 7. The normality condition

REMARK. We can start from (6.14)~(6.15) to formulate the problem (6.1)~(6.3). Assume that (6.14)~(6.15) has a solution $(u,\sigma,\hat{u}^p)$. We introduce the sets

$$E = \{\, t \in I;\ |\sigma| < G(\hat{u}^p)\} \qquad P = \{\, t \in I; |\sigma| = G(\hat{u}^p)\}.$$

The condition (6.15) then implies that

(6.16) $\dot{u} - \frac{1}{k}\dot{\sigma} = 0, \quad \dot{\hat{u}}^p = 0$ on E,

(6.17) $\dot{u} - \frac{1}{k}\dot{\sigma} = \frac{1}{(G')^2}\dot{\sigma}, \quad \dot{\hat{u}}^p = \frac{1}{G'}\frac{\sigma}{|\sigma|}\dot{\sigma}, \quad \sigma\dot{\sigma} \geq 0$ a.e. P.

PROOF. If $t \in E$ then (6.16) is obvious, since $(\sigma, \hat{u}^p)(t)$ is an inner point of B_G. Hence $\hat{u}^p = 0$ holds until the time, say $t = t_0$, at which the point $(\sigma, 0)$ reaches to the boundary of B_G. For $t \in P$ in a small neighbourhood of t_0, there is a function λ $(\lambda \geq 0)$ such that

$$\begin{pmatrix} \dot{u} - \frac{1}{k}\dot{\sigma} \\ -\dot{\hat{u}}^p \end{pmatrix} = \lambda \begin{pmatrix} \frac{\sigma}{|\sigma|} \\ -G'(\hat{u}^p) \end{pmatrix},$$

since (6.15) implies that the vector on the left side is parallel to the outward normal to B at the boundary point $(\sigma, \hat{u}^p)$. Now σ and $\hat{u}^p$ belong to $C^1_+(I)$ by the assumption. Hence, considering the limits of two finite difference quotients (see page 35) we have

$$\frac{d|\sigma|}{dt} = G'(\hat{u}^p)\dot{\hat{u}}^p \qquad \text{a.e. } P.$$

The function λ is determined as follows :

$$\lambda = \frac{1}{(G')^2}\frac{d|\sigma|}{dt} = \frac{1}{(G')^2}\frac{\sigma}{|\sigma|}\dot{\sigma}$$

This proves that (6.16)$\sim$(6.17) hold in a certain neighbourhood of t_0. Since $\sigma\dot{\sigma} \geq 0$ implies that $\hat{u}^p$ is monotonically increasing during the plastic deformation, (6.16)$\sim$(6.17) hold in the whole interval I.

We next define u^p by

$$\dot{u}^p = \dot{u} - \frac{1}{k}\dot{\sigma} \qquad (u^p(0) = 0),$$

and introduce $\bar{u}^p$ and $H(\bar{u}^p)$ by $\bar{u}^p(t) = \int_0^t |\dot{u}^p| dt$ and $H(\bar{u}^p) = G(\hat{u}^p)$.

Note that the variables $\bar{u}^p$ and $\hat{u}^p$ are connected by the relation

$$\bar{u}^p = \int_0^{\hat{u}^p} \frac{1}{G'(s)} \, ds.$$

Hence we finally have the desired equations

$$\dot{u} - \frac{1}{k}\dot{\sigma} = 0, \quad \dot{u}^p = 0 \qquad \text{on } E : \ |\sigma| < H(\bar{u}^p)$$

$$\dot{u} - \frac{1}{k}\dot{\sigma} = \frac{1}{g(\sigma)}\dot{\sigma}, \quad \dot{u}^p = \frac{\dot{\sigma}}{g(\sigma)}, \quad \sigma\dot{\sigma} \geq 0 \qquad \text{a.e. } P : \ |\sigma| = H(\bar{u}^p),$$

where $g(\sigma) = H'(H^{-1}(|\sigma|))$.

CHAPTER 3

ELASTIC-PLASTIC VIBRATION OF A SPRING-MASS SYSTEM WITH MULTIPLE DEGREES OF FREEDOM

3.1. Equation of motion (kinematic hardening problem)

In this chapter we consider an initial value problem of a spring-mass system with multiple degrees of freedom as formulated in Chapter 1. The main purpose is to show the existence of a classical solution, and to do this we apply the idea introduced in Chapter 2 . For the multiple system there is a special difficulty for our method, namely, that some points may yield or unload at the same time. Nevertheless, our method is still valid. In fact, all results obtained for the single mass system can be extended formally to the multiple case.

Let ρ_i and u_i ($i = 1, 2, \ldots, N$) be the mass and displacement of the i-th mass point. Let k_i and ξ_i be the elastic constant and plasticity factor of the i-th spring. The strain of the i-th point is denoted by $U_i = u_i - u_{i-1}$ ($u_0 = 0$). The equation of motion of this system is then written as

$$(1.1) \qquad \rho_i \ddot{u}_i + \sigma_i - \sigma_{i+1} = b_i(t) \qquad i = 1 \sim N,$$

where

$$(1.2) \qquad \begin{cases} \dot{\sigma}_i = k_i \dot{U}_i, \quad \dot{\alpha}_i = 0 & \text{if } |\sigma_i - \alpha_i| < z_0 \\ \dot{\sigma}_i = (1 - \xi_i) k_i \dot{U}_i, \quad \dot{\alpha}_i = \dot{\sigma}_i & \text{if } |\sigma_i - \alpha_i| = z_0 \text{ and } (\sigma_i - \alpha_i)\,\dot{\sigma}_i \geq 0, \end{cases}$$

and $\sigma_{N+1} = 0$. The first (or second) condition of (1.2) is called the elastic (or plastic) stress-strain relation, which must be understood as follows : We define the elastic (or plastic) zone E_i (or P_i^+ and P_i^-) in (U_i, σ_i) plane (see Fig. 8) :

$$E_i = \{ (U_i, \sigma_i) \; ; \; (1 - \xi_i)k_iU_i - \xi_i z_0 < \sigma_i < (1 - \xi_i)k_iU_i + \xi_i z_0 \}$$

$$P_i^{\pm} = \{ (U_i, \sigma_i) \; ; \; \sigma_i \mp z_0 = (1 - \xi_i)k_i(U_i \mp \frac{1}{k_i} z_0)\}.$$

We say that the i-th spring is elastic if $(U_i, \sigma_i) \in E_i$ and plastic if $(U_i, \sigma_i) \in P_i^+ \cup P_i^-$.

DEFINITION. The stress-strain relation is admissible if the following conditions (a), (b) and (c) are fulfilled for the solution of (1.1) :

(a) σ_i and U_i belong to $C_+^1(I)$ on t, and the point (U_i, σ_i) is always included in $E_i \cup P_i^+ \cup P_i^-$.

(b) The relation $\dot{\sigma}_i = k_i\dot{U}_i$ (or $\dot{\sigma}_i = (1 - \xi_i)k_i\dot{U}_i$) holds in E_i (or on $P_i^{\pm}$).

(c) If $(U_i, \sigma_i) \in P_i^+$ (or $\in P_i^-$) then $\dot{U}_i \geq 0$ (or ≤ 0).

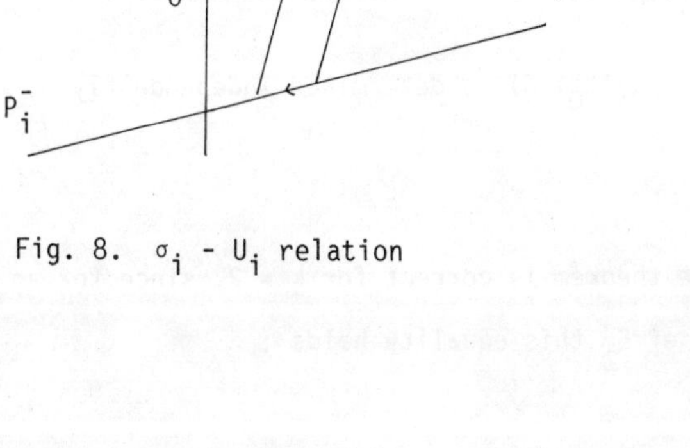

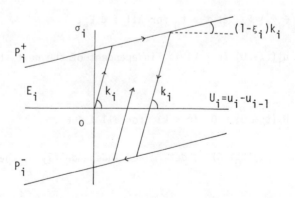

Fig. 8. $\sigma_i - U_i$ relation

Our first problem is to find an admissible stress-strain relation for each spring under which the solution is sought. We assume that $u_i(0) = \sigma_i(0) = 0$ and $\dot{u}_i(0) = a_i$ (= given constant) as the initial condition.

Naturally the elastic stress-strain relation is used until $|\sigma_i| = z_0$ is satisfied for some spring. Assume that at $t = t_0$ this is satisfied by the springs of a set E_1 and that others are still elastic after t_0. If $\dot{U}_i(t\ )\neq 0$, or if $\ddot{U}_i(t\ ) \neq 0$ even when $\dot{U}_i(t_0) = 0$,the next state of the springs of E_1 is determined uniquely as in the single system. Assume that the first and the second derivatives of U_i vanish at $t = t_0$ for all the springs of $E_2 \subset E_1$ and that the stress-strain relations of the springs in $E_1 - E_2$ are already determined . We want to determine the next state of each springs of E_2. The following theorem is essential.

THEOREM 3.1. Let E_k be a set of springs. Assume that the next state of the springs in $E_1 - E_k$ ($k \geq 2$) is already determined independently of the choice of the next state of E_k and, moreover, that the following conditions are satisfied :

(1) $(U_i, \sigma_i) \in P_i^+ \cup P_i^-$ at $t = t_0$ for all $i \in E_k$,

(2) $d^r/dt^r\ u_i(t_0 + 0)$ $(r \leq k)$ is independent of the next state of E_k for all i,

(3) $d^r/dt^r\ U_i(t_0 + 0) = 0$ $(r \leq k)$ for all $i \in E_k$.

Then $d^{k+1}/dt^{k+1}\ u_i(t_0 + 0)$ is determined independently of the next state of E_k for all i.

PROOF. The theorem is correct for $k = 2$, since for an arbitrary choice of the next state of E_2 this equality holds :

$$\rho_i \frac{d^3}{dt^3} u_i + \dot{\sigma}_i - \dot{\sigma}_{i+1} = \dot{b}_i ,$$

and $\dot{\sigma}_j(t_0+0) = [K]_j \dot{U}_j(t_0+0) = 0$ for $j \in E_2$, and $\dot{\sigma}_j(t_0+0) = [K]_j \dot{U}_j(t_0+0)$ for $j \notin E_2$ independently of the next state of the springs in E_2, where $[K]_j$ denotes a definite constant. This argument is valid for any k. To show this, we solve (1.1) by assuming an arbitrary state for E_k. For $t > t_0$ we have

$$\rho_i \frac{d^{k+1}}{dt^{k+1}} u_i + \frac{d^{k-1}}{dt^{k-1}} (\sigma_i - \sigma_{i+1}) = \frac{d^{k-1}}{dt^{k-1}} b_i .$$

Here we have

$$\frac{d^{k-1}}{dt^{k-1}} \sigma_j(t_0+ 0) = [K]_j \frac{d^{k-1}}{dt^{k-1}} U_j(t_0+ 0) = 0$$

for $j \in E_k$ and

$$\frac{d^{k-1}}{dt^{k-1}} \sigma_j(t_0+ 0) = [K]_j \frac{d^{k-1}}{dt^{k-1}} U_j(t_0+ 0)$$

for $j \notin E_k$ independently of the next state of E_k, where $[K]_j$ denotes a definite constant. Therefore, $d^{k-1}/dt^{k-1}\ \sigma_j(t_0+0)$ and hence $d^{k+1}/dt^{k+1}\ u_j(t_0+0)$ are independent of the next state of E_k. This completes the proof.

An admissible stress-strain relation just after t_0 can be determined by this theorem for all the springs. To see this, we first note that the value $d^3/dt^3\ u_i(t_0+0)$ is independent of the next state of E_2. Hence the sign of this value can determine the next state of the spring $i \in E_2$ if it is not zero. If it should vanish, then we can check the sign of the fourth derivative, since its value is independent of the next state of E_3. We repeat this argument until $d^k/dt^k\ U_i(t_0+ 0) \neq 0$ is satisfied for a certain finite k. If this repetition does not terminate, then clearly $\dot{U}_i(t) = 0\ (t \geq t_0)$ for such spring. In any case, the next state can be determined so that the solution is continuated beyond $t = t_0$.

The same argument is valid when unloading may occur, or even when the second yielding or unloading occurs. Thus we can continuate the solution under an admissible stress-strain relation. Since the boundedness of these solutions is assured always by the energy inequality which is derived in the next section, there is a unique solution of our problem in any time interval.

3.2. Energy inequalities

We say that the mass system is in stage(m) ($m = (m_1, m_2, \ldots, m_N)$) if the i-th spring is in stage(m_i). By $U_i^{(j)} = u_i^{(j)} - u_{i-1}^{(j)}$ we denote the strain when the i-th spring enters into stage(j). The following theorems correspond to Theorems 2.4 and 2.5 for the single system :

THEOREM 3.2. At stage(m) the equation of motion of the multiple system is expressed as follows.

$$(2.1)\qquad \rho\ddot{u}_i + [\, k_i U_i - \xi_i k_i \sum_{j=0}^{m_i} (-1)^j (U_i - U_i^{(j)})\,] - [\, k_{i+1} U_{i+1} - \xi_{i+1} k_{i+1} \sum_{j=0}^{m_{i+1}} (-1)^j (U_{i+1} - U_{i+1}^{(j)})\,] = b_i \qquad (i = 1 \sim N,\ k_{N+1} = 0).$$

Since this theorem is proved in just the same way as in Theorem 2.4, we omit it. Let E_m be defined by

$$E_m(t) = \frac{1}{2} \sum_{i=1}^{N} [\rho_i \dot{u}_i^2 + k_i U_i^2 - \xi_i k_i \sum_{j=0}^{m_i} (-1)^j (U_i - U_i^{(j)})^2].$$

THEOREM 3.3. The following equality holds in stage(m) :

$$(2.2)\qquad E_m(t) = \frac{1}{2} \sum_{i=1}^{N} \rho_i \dot{u}_i^2(0) + \sum_{i=1}^{N} \int_0^t b_i \dot{u}_i \, dt.$$

PROOF. Assume that the stage(m) begins from $t = t_0$. Also assume that the springs $i = i_1, i_2, \ldots, i_q$ become plastic or elastic at $t = t_0$ and that the others remain unchanged. Multiply both sides of (2.1) by $\dot{u}_i$, sum up on i, and integrate the resulting equation from t_0 to t in stage(m). Then we have

$$E_m(t) = \frac{1}{2}\sum_{i=1}^{N}[\rho_i\dot{u}_i^2(t_0) + k_iU_i^2(t_0) - \xi_ik_i\sum_{j=0}^{m_i}(-1)^j(U_i(t_0) - U_i^{(j)})^2] + \sum_{i=1}^{N}\int_{t_0}^{t} b_i\dot{u}_i\,dt,$$

since the following identity holds :

$$\sum_{i=1}^{N}[\xi_ik_i\sum_{j=0}^{m_i}(-1)^j(U_i - U_i^{(j)}) - \xi_{i+1}k_{i+1}\sum_{j=0}^{m_{i+1}}(-1)^j(U_{i+1} - U_{i+1}^{(j)})]\,\dot{u}_i$$

$$= \frac{1}{2}\sum_{i=1}^{N}[\xi_ik_i\sum_{j=0}^{m_i}(-1)^j(U_i - U_i^{(j)})\,]_t^2.$$

Let the preceding stage be stage(m'). Then it is clear that

$$E_m(t) = E_{m'}(t_0) + \sum_{i=1}^{N}\int_{t_0}^{t} b_i\dot{u}_i\,dt,$$

since $U_i(t_0) = U_i^{(m_i)}$ for $i = i_1, i_2, \ldots, i_q$. Hence (2.2) follows.

THEOREM 3.4. Assume that the solution is continuated up to T_0. Then the following estimates hold for any $t \le T_0$:

$$(2.3)\qquad \frac{1}{2}\sum_{i=1}^{N}[\rho_i\dot{u}_i^2 + \frac{(1-\xi_i)}{2}k_iU_i^2](t) \le [C + \frac{1}{2}\int_0^t(\sum_{i=1}^{N}\frac{2}{\rho_i}b_i^2)^{\frac{1}{2}}ds]^2$$

$$(2.4)\qquad \frac{1}{2}\sum_{i=1}^{N}[\rho_i\ddot{u}_i^2 + (1-\xi_i)k_i\dot{U}_i^2](t) \le [\bar{C} + \frac{1}{2}\int_0^t(\sum_{i=1}^{N}\frac{2}{\rho_i}\dot{b}_i^2)^{\frac{1}{2}}ds]^2,$$

where C and $\bar{C}$ are constants depending only on the given data.

PROOF. Set

$$S_i = \frac{k_i}{2} U_i^2 - \frac{\xi_i k_i}{2} \sum_{j=0}^{m_i} (-1)^j (U_i - U_i^{(j)})^2.$$

As in the proof of Theorem 2.6 of Chapter 2, we have

$$S_i \geq \frac{1}{4}(1 - \xi_i)k_i U_i^2 - \frac{\xi_i(1+\xi_i)k_i}{2(1-\xi_i)} (U_i^{(0)})^2,$$

where $U_i^{(0)}$ is the strain when the first yielding occurs : $U_i^{(0)} = 1/k_i\ z_0$.

Therefore we have by Theorem 3.3

$$(2.5) \quad \frac{1}{2} \sum_{i=1}^{N} [\ \rho_i \dot{u}_i^2 + \frac{(1-\xi_i)}{2} k_i U_i^2](t)$$
$$\leq \frac{1}{2} \sum_{i=1}^{N} [\ \rho_i \dot{u}_i^2(0) + \frac{\xi_i(1+\xi_i)k_i}{(1-\xi_i)} (U_i^{(0)})^2] + \sum_{i=1}^{N} \int_0^t |b_i \dot{u}_i| dt.$$

Let g(t) and h(t) be the vectors defined by

$$g(t) = (\sqrt{\frac{\rho_i}{2}}\ \dot{u}_i, \frac{\sqrt{(1-\xi_i)k_i}}{2} \quad U_i\ )$$

$$h(t) = (\ \sqrt{\frac{2}{\rho_i}}\ \ |b_i|,\ 0\ ).$$

Then the identity (2.5) is written as

$$(2.6) \qquad (g,\ g)_{R^{2N}} \leq C^2 + \int_0^t |(h,\ g)_{R^{2N}}|\ ds,$$

where C^2 is the first term of the right side of (2.5). Let z(t) be the right side of (2.6). Then

$$[z'(t)]^2 \leq \|h(t)\|^2\ \|g(t)\|^2 \leq \|h(t)\|^2\ z(t),$$

so that $z' \leq \|h\|\ z^{1/2}$. Hence z must satisfy

$$z(t)^{\frac{1}{2}} \leq C + \frac{1}{2} \int_0^t \|h(s)\| ds,$$

from which (2.3) follows. To prove (2.4) we first note that by (2.1)

$$(2.7) \qquad \frac{1}{2} \sum_{i=1}^{N} [\ \rho_i (\ddot{u}_i)^2 + k_i (1 - \xi_i \sum_{j=0}^{m_i} (-1)^j)(\dot{U}_i)^2]_t = \sum_{i=1}^{N} \dot{b}_i \ddot{u}_i\ .$$

Assume now that the stage(m) begins from $t = t_0$ and that

$$i = i_1, \ldots, i_p \qquad \text{become plastic,}$$
$$i = i_{p+1}, \ldots, i_q \qquad \text{become elastic}$$

at $t = t_0$, and other springs keep the same state. If the preceding stage is denoted by m' , then it is clear that

$$m_i = \begin{cases} m_i' & i \neq i_1, \ldots, i_q \\ m_i' + 1 & i = i_1, \ldots, i_q \end{cases}$$

Define

$$\bar{E}_m(t) = \frac{1}{2} \sum_{i=1}^{N} [\rho_i \ddot{u}_i^2 + k_i(1 - \xi_i \sum_{j=0}^{m_i} (-1)^j)\dot{U}_i^2](t).$$

Then by (2.7) we have

$$\text{(2.8)} \qquad \bar{E}_m(t) - \bar{E}_{m'}(t') = \bar{E}_m(t_0) - \bar{E}_{m'}(t_0) + \sum_{i=1}^{N} \int_{t'}^{t} \dot{b}_i \ddot{u}_i \, dt$$

for any t' and t near t_0 ($t' < t_0 < t$), and

$$\bar{E}_m(t_0) - \bar{E}_{m'}(t_0) = \frac{1}{2} \sum_{i=1}^{N} [\rho_i \ddot{u}_i^2(t_0) + k_i(1 - \xi_i \sum_{j=0}^{m_i} (-1)^j)\dot{U}_i^2(t_0)]$$
$$- \frac{1}{2} \sum_{i=1}^{N} [\rho_i \ddot{u}_i^2(t_0) + k_i(1 - \xi_i \sum_{j=0}^{m_i'} (-1)^j)\dot{U}_i^2(t_0)].$$

Since $\ddot{u}_i$ is continuous, the right side of this identity is equal to the sum of

$$- \frac{1}{2} k_i \xi_i (-1)^{m_i} \dot{U}_i^2(t_0) \qquad (\leq 0)$$

for $i = i_1, \ldots, i_p$. Here we used the facts that m_i is even for $i = i_1, \ldots, i_p$ and that $\dot{U}_i(t_0) = 0$ for $i = i_{p+1}, \ldots, i_q$. Repeating this argument, we finally have

$$\bar{E}_m(t) \leq \frac{1}{2} \sum_{i=1}^{N} [\rho_i \ddot{u}_i^2(0) + k_i \dot{U}_i^2(0)] + \sum_{i=1}^{N} \int_{0}^{t} \dot{b}_i \ddot{u}_i \, dt,$$

from which (2.4) follows as before.

3.3. Behaviour of the solution as $t \to \infty$

Since the boundedness of the solution is now assured, there exists a unique solution of our problem in $(0,\infty)$ provided $b_i(t)$ are defined in this interval. In fact, if the number of the state change is finite in any finite time interval, then this is clear. If it is infinite, we have to be careful in expressing the equation of motion and $E_m(t)$. In this case we exclude, as we did for the single mass system, those terms for which $U_i^{(j)} = U_i^{(j+1)}$ holds, since these terms cancel each other. Then the $U_i^{(j)}$ in (2.1) consists of those which contribute to form a hysteresis loop and, as the result, m_i is finite in a finite time interval, since the number of such loops must be finite in a finite time interval by the energy inequality.

Now corresponding to Theorem 2.8 we have

THEOREM 3.5. If one of the following conditions is satisfied, then the elastic-plastic vibration converges to an elastic vibration as $t \to \infty$.

(a) $\int_0^\infty \left(\sum_{i=1}^N b_i^2\right)^{\frac{1}{2}} dt < \infty$

(b) $\sup_{i,t} |b_i(t)| < \infty$ and $\int_0^\infty \left(\sum_{i=1}^N \dot{b}_i^2\right)^{\frac{1}{2}} dt < \infty$.

PROOF. Suppose that the system is in stage(m), and set

$$E_i^*(t) = -\frac{k_i\xi_i}{2}\sum_{j=1}^{m_i}(-1)^j(U - U^{(j)})^2.$$

It suffices to show that this quantity is finite as $t \to \infty$. We first see

$$\frac{k_i}{2}U_i^2 - \frac{k_i\xi_i}{2}(U_i - U_i^{(0)})^2 \geq \frac{1}{4}(1-\xi_i)k_iU_i^2 - \frac{\xi_i(1+\xi_i)k_i}{2(1-\xi_i)}(U_i^{(0)})^2.$$

Therefore, by Theorem 3.3, there is a constsnt C such that

$$\frac{1}{2}\sum_{i=1}^{N} [\rho_i \dot{u}_i + \frac{1}{2}(1 - \xi_i)k_i U_i^2 + E_i^*(t)] \leq C + \sum_{i=1}^{N}\int_0^t b_i \dot{u}_i \, dt.$$

Hence the theorem is proved in the same way as in the single system.

3.4. A weak form

As in the single system, the present problem can be represented by a weak form including an inequality. Let u, σ and α be N-dimensional vector functions whose components belong to the class $C_+^1(I)$. Let K be the set of N-dimensional vector functions which are within the z_0-neighbourhood of α :

$$K = K_\alpha = \{ \tau \in C_+^1(I) \; ; \; \operatorname*{Max}_i |\tau_i - \alpha_i| \leq z_0 \quad \text{for all } t \in I \}.$$

THEOREM 3.6. The initial value problem of the multiple mass system is equivalent to the following problem : Seek $(u, \sigma, \alpha) \in C_+^1(I)$ which satisfies for all $t \in I$

$$(4.1) \qquad \rho_i \ddot{u}_i + \sigma_i - \sigma_{i+1} = b_i \qquad \text{for } i = 1 \sim N$$

$$(4.2) \qquad \sum_{i=1}^{N} (\dot{\sigma}_i - k_i \dot{U}_i)(\tau_i - \sigma_i) \geq 0 \qquad \text{for any } \tau \in K$$

$$(4.3) \qquad \dot{\alpha}_i = (1 - \frac{1}{\xi_i})(\dot{\sigma}_i - k_i \dot{U}_i)$$

with $\sigma \in K$, $u(0) = 0$, $\dot{u}(0) = a$, $\sigma(0) = \sigma_{N+1} = \alpha(0) = 0$.

PROOF. Let $\{u_i\}$ be the solution of the previous problem. Take an arbitrary $t \in I$ and assume that, in a certain time interval $[t, t + \delta)$ ($\delta > 0$), the $\dot{\sigma}$ - $\dot{U}$ relation is given by

$$\dot{\sigma}_i = k_i\dot{U}_i \qquad \text{for } i = i_1, \ldots, i_r$$
$$\dot{\sigma}_i = (1 - \xi_i)k_i\dot{U}_i \qquad \text{for } i = i_{r+1}, \ldots, i_N .$$

If the spring i is elastic (or plastic) in this interval, then it is clear that α_i satisfies (4.3) since $\dot{\alpha}_i = 0$ (or $= \dot{\sigma}_i$). Also, in this interval,

$$\sum_{i=1}^{N} (\dot{\sigma}_i - k_i\dot{U}_i)(\tau_i - \sigma_i) = - \sum_{i=i_{r+1}}^{i_N} \xi_i k_i \dot{U}_i (\tau_i - \sigma_i).$$

Now if the spring i is plastic, we have

$$\sigma_i = \alpha_i + z_0 \qquad \text{if } \dot{U}_i > 0$$

$$\sigma_i = \alpha_i - z_0 \qquad \text{if } \dot{U}_i < 0,$$

which imply

$$\dot{U}_i(\tau_i - \alpha_i) \leq 0 \qquad \text{for all } \tau_i \ ; \ |\tau_i - \alpha_i| \leq z_0,$$

and hence (4.2) follows. The uniqueness of the solution of (4.1) $\sim$ (4.2) can be proved in the same way as in the single system.

3.5. Isotropic hardening problem

The governing equations are

(5.1) $$\rho_i\ddot{u}_i + \sigma_i - \sigma_{i+1} = b_i \qquad i = 1 \sim N,$$

(5.2) $$\begin{cases} \dot{\sigma}_i = k_i\dot{U}_i, \quad \dot{U}_i^p = 0 & \text{if } |\sigma_i| < H(\bar{U}_i^p) \\ \dot{\sigma}_i = \left(k_i - \dfrac{k_i^2}{g(\sigma_i)+k_i}\right)\dot{U}_i, \quad \dot{U}_i^p = \dfrac{1}{g(\sigma_i)}\dot{\sigma}_i & \text{if } |\sigma_i| = H(\bar{U}^p) \text{ and } \sigma_i\dot{\sigma}_i \geq 0 \end{cases}$$

where

$$U_i = u_i - u_{i-1} \qquad (u_0 = 0)$$

$$\bar{U}_i^p = \int_0^t |\dot{U}_i^p| \, dt$$

$$g(\sigma_i) = H'(H^{-1}(|\sigma_i|)), \qquad H(0) = z_0 ,$$

$$u_i(0) = \sigma_i(0) = U_i^p(0) = \sigma_{N+1} = 0, \ \dot{u}_i(0) = a_i \quad (= \text{given}).$$

The existence and uniqueness of the solution of this problem can be proved in the same way as in the single system. Here is the outline. Assume that at $t = t_0$ some springs satisfies $|\sigma_i| = z_0$. The next state of these springs is determined if the first or the second derivative of $\dot{U}_i$ does not vanish at $t = t_0$. For those springs for which $\dot{U}_i(t_0) = \ddot{U}_i(t_0) = 0$ hold, we can employ the next theorem, which corresponds to Theorem 3.1.

THEOREM 3.7. Let E_k $(k \geq 2)$ be a set of springs satisfying the following conditions : The next state of the springs which are not included in E_k is already determined independently of the next state of E_k , and

(1) $|\sigma_i| = H(0)$ for $i \in E_k$

(2) $d^r/dt^r \ u_i(t_0 + 0)$ $(r \leq k)$ is independent of the next state of E_k for all i,

(3) $d^r/dt^r \ U_i(t_0 + 0) = 0$ $(r \leq k)$ for all $i \in E_k$.

Then $d^{k+1}/dt^{k+1} \ u_i(t_0 + 0)$ is determined independently of the next state of E_k for all i.

PROOF. Consider a solution of (5.1) obtained by assuming an arbitrary next state of the springs of E_k . Then if $i \in E_k$, condition (3) implies that

$$\frac{d^r}{dt^r} \sigma_i(t_0 + 0) = 0 \qquad (r \leq k).$$

If $i \notin E_k$, then the above derivatives are determined independently of the next state of E_k. To see this, we first note that this is clear for elastic i since in this case it holds that

$$\frac{d^r}{dt^r}\sigma_i(t_0 + 0) = k_i \frac{d^r}{dt^r} U_i(t_0 + 0).$$

If spring i is plastic, we can use an induction to prove the assertion, since $d^r/dt^r \sigma_i(t_0 + 0)$ is determined by the derivatives of order $\leq r-1$ of σ_i and of order $\leq r$ of U_i at $t = t_0 + 0$. Now we have the identity

$$\rho_i \frac{d^{k+1}}{dt^{k+1}} u_i + \frac{d^{k-1}}{dt^{k-1}} (\sigma_i - \sigma_{i+1}) = \frac{d^{k-1}}{dt^{k-1}} b_i,$$

for $t \geq t_0 + 0$. Hence the theorem follows.

By this theorem we can continuate the solution beyond $t = t_0$. Since the energy estimates are derived as shown below, the situation is identical to that in kinematic hardening, and we obtain a desired solution in I.

THEOREM 3.8. The following estimates hold for all $t \in I$:

$$(5.3) \qquad \frac{1}{2}\sum_{i=1}^{N} [\rho_i \dot{u}_i^2 + \frac{1}{k_i} \sigma_i^2]_0^t + \int_0^t \sigma_i \dot{U}_i^p \, dt = \sum_{i=1}^{N} \int_0^t b_i \dot{u}_i \, dt$$

$$(5.4) \qquad \frac{1}{2}\sum_{i=1}^{N} [\rho_i \ddot{u}_i^2 + \frac{1}{k_i} \dot{\sigma}_i^2 + g(\sigma_i)(\dot{U}_i^p)^2]_0^t \leq \sum_{i=1}^{N} \int_0^t \dot{b}_i \ddot{u}_i \, dt.$$

PROOF. For all $t \in I$ the next relations holds :

$$\dot{U}_i - \frac{1}{k_i}\dot{\sigma}_i = 0, \quad \dot{U}_i^p = 0 \qquad \text{for elastic spring,}$$

$$\dot{U}_i - \frac{1}{k_i}\dot{\sigma}_i = \frac{1}{g(\sigma_i)}\dot{\sigma}_i = \dot{U}_i^p \qquad \text{for plastic spring.}$$

Hence for all $t \in I$ we have

$$(5.5) \qquad \sum_{i=1}^{N} \sigma_i(\dot{U}_i - \frac{1}{k_i}\dot{\sigma}_i - \dot{U}_i^p) = 0.$$

Here we have

$$\sum_{i=1}^{N} \sigma_i \dot{U}_i = \sum_{i=1}^{N} (\sigma_i - \sigma_{i+1})\dot{u}_i = \sum_{i=1}^{N} (b_i \dot{u}_i - \rho_i \ddot{u}_i \dot{u}_i).$$

Substituting this into (5.5) and integrating on t, we have (5.3). The estimate (5.4) is obtained the same way as in the single mass system (see proof of Theorem 2.12).

Finally, we shall express the problem (5.1)~(5.2) in a weak form. We introduce a new parameter $\hat{U}_i^p$ defined by

$$\hat{U}_i^p = \int_0^{\bar{U}_i^p} \sqrt{H'(t)}\, dt.$$

Let K_G be the set defined in Sec. 6 of Chap. 2. The next theorem is proved in just the same way in the single mass system.

THEOREM 3.9. The problem (5.1)~(5.2) is equivalent to the following problem : Seek $(u\ ,\ \sigma, \hat{U}^p) \in C_+^1(I)$ which satisfies for all $t \in I$

$$(5.6) \qquad \rho_i \ddot{u}_i + \sigma_i - \sigma_{i+1} = b_i \qquad i = 1 \sim N,$$

$$(5.7) \qquad \sum_{i=1}^{N} [(\dot{U}_i - \frac{1}{k_i}\dot{\sigma}_i)(\tau_i - \sigma_i) - \dot{\hat{U}}_i^p(\zeta_i - \hat{U}_i^p)] \le 0$$

$$\text{for all } (\tau_i, \zeta_i) \in K_G$$

with $(\sigma_i, \hat{U}_i^p) \in K_G$, $u_i(0) = \sigma_i(0) = \hat{U}_i^p(0) = 0$, $\dot{u}_i(0) = a_i$.

CHAPTER 4

QUASI-STATIC PROBLEMS OF A SPRING-MASS SYSTEM WITH MULTIPLE DEGREES OF FREEDOM

4.1 Continuation of the solution (kinematic hardening problem)

The quasi-static problem of multiple mass system formulated in Chapter 1 was as follows: Assume kinematic hardening. Let u_i be the displacement of the i-th mass point. Let U_i, σ_i, and α_i be the strain , stress, and the parameter to represent the center of the yield surface of the i-th spring. These quantities must satisfy the following relations :

$$\sigma_i - \sigma_{i+1} = b_i(t) \qquad i = 1 \sim N, \tag{1.1}$$

$$\begin{cases} \dot{\sigma}_i = k_i\dot{U}_i, \ \dot{\alpha}_i = 0 & \text{if } |\sigma_i - \alpha_i| < z_0 \text{ or } |\sigma_i - \alpha_i| = z_0 \text{ and} \\ & \qquad (\sigma_i - \alpha_i)\dot{U}_i < 0 \\ \dot{\sigma}_i = (1 - \xi_i)k_i\dot{U}_i, \ \dot{\alpha}_i = \dot{\sigma}_i & \text{if } |\sigma_i - \alpha_i| = z_0 \text{ and} \\ & \qquad (\sigma_i - \alpha_i)\dot{U}_i \geq 0, \end{cases} \tag{1.2}$$

where $U_i = u_i - u_{i-1}$. The function $b_i(t)$ is continuous, piecewise analytic, and $b_i(0) = 0$. $k_i(> 0)$, z_0 and $\xi_i (0 \leq \xi_i < 1)$ are given constants. We want a $C^1_+(I)$ or, if possible, piecewise analytic solution (u,σ,α) under the initial-boundary conditions $u(0) = \sigma(0) = \alpha(0) = 0$ and $u_0 = u_{N+1} = 0$.

In a neighborhood of t = 0, all the springs are clearly elastic ; that is,

$$\dot{\sigma}_i = k_i\dot{U}_i \ , \quad \dot{\alpha}_i = 0 \qquad i = 1 \sim N+1$$

so that equation (1.1) takes the form

$$- k_i u_{i-1} + (k_i + k_{i+1})u_i - k_{i+1}u_{i+1} = b_i(t) \qquad i = 1 \sim N,$$

where $u_{N+1} = 0$. Hence our problem has a unique solution until $t = t_0$ at which some spring satisfies $|\sigma_i| = z_0$. Let E and E_0 be, respectively, the set of all the springs (1, 2,..., N+1) and its subset consisting of the springs for which $|\sigma_i| = z_0$ is satisfied at $t = t_0$. The springs in $E - E_0$ are still elastic after t_0. The springs in E_0, however, may or may not yield after t_0. In the dynamic problem, we could use the inertia term to forecast the sign of the stress velocity. In the quasi-static problems, such a term does not exist and, in general, we cannot expect even the continuity of the strain velocity $\dot{U}_i$ across t_0. Let the number of the springs in E_0 be M. The number of possible combinations of the states after t_0 is then 2^M. Thus our first questions are

(1) Does exist a combination of the states which satisfies (1.2) ?

(2) Is this combination unique ?

We shall show below that both answers are yes and that in fact only one in 2^M cases is realized.

First we assume that there is a desirable combination of the states, and then we try to guess the sign of $\dot{U}_i(t_0 + 0)$ when the problem is solved according to this combination . To do this, consider the problem of finding the solution (u^0,σ^0) of the following system of equations :

(1.3) $$\sigma_i^0 - \sigma_{i+1}^0 = \dot{b}_i(t_0 + 0) \qquad i = 1 \sim N,$$

where σ_i^0 are determined from the following equations.

(1.4) $$\sigma_i^0 = k_i U_i^0 \qquad \text{for } E - E_0,$$

and for E_0

$$(1.5) \begin{cases} \sigma_i^0 = k_i U_i^0 & \text{in } D_- = \{ u^0 \in R^N \ ; \ (\sigma_i(t_0) - \alpha_i(t_0))U_i^0 < 0 \} \\ \sigma_i^0 = (1 - \xi_i)k_i U_i^0 & \text{in } D_+ = \{ u^0 \in R^N \ ; \ (\sigma_i(t_0) - \alpha_i(t_0))U_i^0 \geq 0 \}, \end{cases}$$

where $U_i^0 = u_i^0 - u_{i-1}^0$ and $u_0^0 = u_{N+1}^0 = 0$. Note that if there is a solution of (1.1) $\sim$ (1.2) after t_0, then $(\dot{u},\dot{\sigma})(t_0 + 0)$ of this solution satisfies (1.3)$\sim$(1.5).

THEOREM 4.1. There exists a unique solution (u^0, σ^0) of (1.3)$\sim$(1.5). This solution minimizes the functional

$$F_1(u^0) = \frac{1}{2}(\sigma^0, U^0) - (\dot{b}(t_0 + 0), u^0)$$

under the conditions (1.4)$\sim$(1.5).

PROOF. We first show that

$$F^i \equiv \frac{1}{2}\sigma_i^0 U_i^0 - \dot{b}_i(t_0 + 0)u_i^0$$

is a C^1-class function of u^0. The continuity is clear since σ_i^0 is continuous. The smoothness is also clear for $i \in E - E_0$. If $i \in E_0$, since $\sigma_i^0 U_i^0 = k_i (U_i^0)^2$ holds in D_-, we have in D_-

$$(1.6) \qquad \partial F^i/\partial u_p^0 = \sigma_i^0 \partial U_i^0/\partial u_p^0 - \dot{b}_i(t_0 + 0)\delta_{ip},$$

where δ_{ip} is Kronecker's delta, and in D_+,

$$(1.7) \qquad \partial F^i/\partial u_p^0 = (1 - \xi_i)k_i U_i \, \partial U_i^0/\partial u_p^0 - \dot{b}_i(t_0 + 0)\delta_{ip}$$

$$= \sigma_i^0 \partial U_i^0/\partial u_p^0 - \dot{b}_i(t_0 + 0)\delta_{ip}.$$

Hence F^i, and so F_1, are of C^1-class. Since $0 \leq \xi_i < 1$ and $F_1(u^0) \to \infty$ as $\|u^0\|$

$\to \infty$, $F_1(u^0)$ is bounded below and has a minimum point which is, at the same time, a stationary point. Therefore, if u^0 is a stationary point of F_1 then $\sigma^0 = \sigma^0(u^0)$ must satisfy

$$\frac{\partial F_1}{\partial u_i^0} = \sigma_i^0 - \sigma_{i+1}^0 - \dot{b}_i(t_0 + 0) = 0.$$

Hence the pair of the minimum point u^0 and σ^0 which is determined by (1.4) $\sim$ (1.5) is a solution. To prove the uniqueness of the solution of (1.3) $\sim$ (1.5), it suffices to show the uniqueness of the stationary point of $F_1(u^0)$. To do this, consider the hyperplane between D_+ and D_- for each $i \in E_0$:

$$X_i = \{ u^0 \in R^N \ ; \ (\sigma_i(t_0) - \alpha_i(t_0)) U_i^0 = 0 \}.$$

Let $\{R_j\}$ be the partition of R^N by these hyperplanes. In each R_j , the state is unique for all the springs and $F_1(u^0)$ is a positive definite quadratic form, so that there is at most one stationary point in it. Now assume that there are two stationary points $\underline{u}^0 \in R_i$ and $u^0 \in R_j$ ($i \neq j$). Consider the line

$$L \ : \underline{u}^0 + t(u^0 - \underline{u}^0) \qquad t \in [0, 1].$$

When t moves from 0 to 1, this line passes through at least two regions of $\{R_j\}$. The function $f(t) = F_1(\underline{u}^0 + t(u^0 - \underline{u}^0))$ is a non-degenerate quadratic form and of C^1-class in [0,1]. Therefore, if $\underline{u}^0$ is a stationary point of $F_1(u^0)$, it is also the minimum point in R_i, and $f(t)$ must be strictly increasing in [0,1]. This contradicts the assumption that u^0 is another stationary point. This completes the proof.

Let (u^0, σ^0) be the solution of (1.3)$\sim$(1.5). The sign of $(\sigma(t_0) - \alpha(t_0)) U_i^0$ is therefore known. We tentatively classify the springs of E_0 into the elastic (E_0^e) and plastic (E_0^p) classes according as this sign is negative and non-negative.

Now consider the following initial value problem set up at $t = t_0$:

(1.8) $$\sigma_i - \sigma_{i+1} = b_i(t) \qquad i = 1 \sim N,$$

(1.9) $$\begin{cases} \dot{\sigma}_i = k_i\dot{U}_i, \quad \dot{\alpha}_i = 0 & \text{for } E - E_0^p \\ \dot{\sigma}_i = (1 - \xi_i)k_i\dot{U}_i, \quad \dot{\alpha}_i = \dot{\sigma}_i & \text{for } E_0^p, \end{cases}$$

where $U_i = u_i - u_{i-1}$ and $(u,\sigma,\alpha)(t_0) = (u,\sigma,\alpha)(t_0 - 0)$, $u_0 = u_{N+1} = 0$.

LEMMA 4.2. The initial value problem (1.8)$\sim$(1.9) has a unique solution (u,σ,α), and $(\dot{u},\dot{\sigma})(t_0 + 0) = (u^0,\sigma^0)$ (= the solution of (1.3)$\sim$(1.5)).

PROOF. Differentiate both sides of (1.8) and denote the resulting equation as (1.8)'. Then the system (1.8)'$\sim$(1.9) becomes a system on $(\dot{u},\dot{\sigma},\dot{\alpha})$. Since ξ_i is less than unity, this problem has a unique solution. The solution of (1.8)$\sim$(1.9) is obtained by integrating this solution. Now the solution (u^0, σ^0) of (1.3)$\sim$(1.5) satisfies (1.8)'$\sim$(1.9) at $t = t_0 + 0$. Since the solution of (1.8)'$\sim$(1.9) is unique, we have the desired equality.

By E_1 we denote the set of points of E_0^p for which the solution of (1.8)$\sim$(1.9) satisfies $\dot{U}_i(t_0 + 0) = 0$. If E_1 is empty, then the points of E_0^e and E_0^p move so as to satisfy the subsidiary conditions of (1.2) during at least a small time interval after t_0. Hence our classification of E_0 is justified, and the solution of (1.1)$\sim$(1.2) is continuated beyond $t = t_0$. If, however, E_1 is not empty, then the state of the springs in E_1 after t_0 is still unknown. In this case, $\dot{U}(t_0 + 0)$ is not enough to determine the next state. Also, it might happen that the state of the points of $E_0 - E_1$ is dependent upon the choice of the next state of E_1. This possibility, however, does not exist.

In fact we have

LEMMA 4.3. Assume that E_1 is not empty. Place some springs of E_1 into E_0^e and solve the problem (1.8)∼(1.9) for this new E_0^p. Then $(\dot{u},\dot{\sigma})(t_0+0)$ is again equal to the solution (u^0,σ^0) of (1.3)∼(1.5). In other words, $(\dot{u},\dot{\sigma})(t_0+0)$ is determined independently of the choice of the next state of E_1.

PROOF. The solution of (1.8)∼(1.9) for the new E_0^p satisfies at $t=t_0+0$

$$\dot{\sigma}_i - \dot{\sigma}_{i+1} = \dot{b}_i(t) \qquad i = 1 \sim N, \tag{1.10}$$

$$\begin{cases} \dot{\sigma}_i = k_i\dot{U}_i & \text{for } E - (\text{new})E_0^p \\ \dot{\sigma}_i = (1-\xi_i)k_i\dot{U}_i & \text{for } (\text{new})E_0^p. \end{cases} \tag{1.11}$$

Since $U_i^0 = 0$ for the springs in E_1, (u^0,σ^0) also satisfies these relations at $t = t_0 + 0$. Since the solution of (1.10)∼(1.11) is unique, the lemma follows.

If E_1 is not empty, we have to guess $\ddot{U}_i(t_0+0)$ to determine the next state of the springs in this set. Hence the above argument must be applied again as follows :

(1) We consider the equation which must be satisfied by $\ddot{u}(t_0+0)$, assuming there is a combination of admissible states of the springs in E_1. The unknown is denoted again by (u^0,σ^0).

(2) The unique solvability of this equation is shown by reducing the problem to a minimization problem.

(3) The springs in E_1 are classified into elastic and plastic ones according to the sign of $(\sigma_i(t_0) - \alpha_i(t_0))U^0$, and a problem corresponding to (1.8)∼(1.9) is solved under this classification. Let the solution be (u,σ,α).

(4) It is proved that $(\ddot{u},\ddot{\sigma})(t_0+0) = (u^0,\sigma^0)$.

(5) It is proved that $(\ddot{u},\ddot{\sigma})(t_0+0)$ is independent of the next state of the springs in E_1 for which $U_i^0 = 0$ holds.

We shall generalize this procedure for later use. Assume there is a set of the springs E_k $(k \geq 1)$. It is empty or, if not, satisfies the following conditions :

For all the springs in E the next state is assigned at $t = t_0$ so as to satisfy the followings : Assume that we solve the equation

$$\sigma_i - \sigma_{i+1} = b_i(t) \qquad i = 1 \sim N$$

for $t > t_0$ according to this σ - U relation. Let (u,σ) be the solution. Then it holds that :

(A) The quantities

$$\frac{d^r}{dt^r}(u,\sigma)(t_0+0) \qquad (\text{ for all } r \leq k\ )$$

are determined independently of the choice of the next state of E_k .

(B) For the spring i in $E_0 - E_k$, there is an n_i $(1\leq n_i\leq k)$ depending on i such that

$$\frac{d^r}{dt^r}U_i(t_0+0) = 0 \quad (1\leq r<n_i), \qquad U_i^{(n_i)} \equiv \frac{d^{n_i}}{dt^{n_i}}U_i(t_0+0) \neq 0$$

and the next state is so assigned that the spring is elastic (or plastic) after t_0 if $(\sigma_i(t_0) - \alpha_i(t_0))U_i^{(n_i)}$ is negative (or positive). The springs in $E - E_0$ are still elastic, of course.

(C) For all the springs in E_k it holds that

$$\frac{d^r}{dt^r}U_i(t_0+0) = 0 \qquad (1\leq r\leq k) .$$

For $k = 1$, the E_1 previously defined satisfies all these conditions. To see this, we first note that the springs in $E - E_0$ are still elastic and those in E_0 are classified according to the sign of $(\sigma_i(t_0) - \alpha_i(t_0))U_i^0$. The problem (1.8)~(1.9) is solved under this stress-strain relation, and Lemma 4.3 holds for the solution (u, σ). Therefore (A) holds. Since E_1 consists of those for which $\dot{U}_i(t_0+0) = 0$ holds, obviously (B) and (C) hold.

Now if E_k is not empty, the state of the springs in this set is still tentative. We determine the next state of the springs in E_k by the following procedure :

(Step 1). We seek the solution (u^0,σ^0) of the system

$$(1.12)\qquad \sigma_i^0 - \sigma_{i+1}^0 = \frac{d^{k+1}}{dt^{k+1}} b_i(t_0+0) \qquad i = 1\sim N,$$

$$(1.13)\qquad \begin{cases} \sigma_i^0 = k_i U_i^0 & \text{for the elastic spring in } E - E_k \\ \sigma_i^0 = (1-\xi_i)k_i U_i^0 & \text{for the plastic spring in } E - E_k, \end{cases}$$

and for those in E_k

$$(1.14)\qquad \begin{cases} \sigma_i^0 = k_i U_i^0 & \text{in } D_- = \{ u^0 \,;\, (\sigma_i(t_0) - \alpha_i(t_0))U_i^0 < 0\} \\ \sigma_i^0 = (1-\xi_i)k_i U_i^0 & \text{in } D_+ = \{ u^0 \,;\, (\sigma_i(t_0) - \alpha_i(t_0))U_i^0 \geq 0\} \end{cases}$$

where $U_i^0 = u_i^0 - u_{i-1}^0$ and $u_0^0 = u_{N+1}^0 = 0$. We then have

LEMMA 4.4. The problem (1.12)~(1.14) has a unique solution which minimizes the functional

$$F_{k+1}(u^0) = \frac{1}{2}(\sigma^0,U^0) - (\frac{d^{k+1}}{dt^{k+1}} b(t_0+0),u^0)$$

under the subsidiary conditions (1.13)~(1.14).

(Step 2). Using the solution (u^0,σ^0) of the problem (1.12)~(1.14), we classify E_k as follows :

$$E_k^e = \{ i \in E_k \ ; \ (\sigma_i(t_0) - \alpha_i(t_o))U_i^0 < 0 \}$$

$$E_k^p = \{ i \in E_k \ ; \ (\sigma_i(t_0) - \alpha_i(t_0))U_i^0 \geq 0 \}.$$

We shall regard the springs in E_k^e (or E_k^p) as elastic (or plastic), and solve the following equation :

$$\text{(1.15)} \qquad \sigma_i - \sigma_{i+1} = b_i(t) \qquad i = 1 \sim N,$$

$$\text{(1.16)} \qquad \begin{cases} \dot{\sigma}_i = k_i \dot{U}_i & \text{for the elastic spring} \\ \dot{\sigma}_i = (1 - \xi_i)k_i\dot{U}_i & \text{for the plastic spring,} \end{cases}$$

where $U_i = u_i - u_{i-1}$, $u_0 = u_{N+1} = 0$ and $(u,\sigma)(t_0) = (u,\sigma)(t_0 - 0)$. Then the following lemma, which corresponds to Lemmas 4.2 and 4.3, holds.

LEMMA 4.5. Let (u,σ) and (u^0,σ^0) be the solutions of (1.15)~(1.16) and of (1.12)~(1.14), respectively. Then it hold that

(a) $\dfrac{d^{k+1}}{dt^{k+1}} (u,\sigma)(t_0 + 0) = (u^0,\sigma^0)$, and that

(b) the value of the left side of (a) is determined independently of the choice of the next state of the springs in E_k^p for which $U_i^0 = 0$ holds.

Now let E_{k+1} be the set of springs in E_k^p such that $U_i^0 = 0$ is satisfied. If E_{k+1} is empty, then our classification of E_k is correct; if it is not, the previous conditions (A),(B) and (C) are satisfied replacing k with k+1. In fact, to prove (A) we note that since $E_{k+1} \subset E_k$, the values of the derivatives of order less than or equal to k are independent of the next state of E_{k+1}

and those of order k + 1 are also independent of it by (b) of Lemma 4.5. Hence (A) holds well. For the springs in $E_0 - E_k$, (B) is clearly satisfied. For the springs in $E_k - E_{k+1}$, the condition of (B) is satisfied by (a) of Lemma 4.5. Hence (B) is satisfied. Finally, (C) is obvious by (a) of Lemma 4.5 and by the definition of E_{k+1}.

(Step 3). If E_{k+1} is empty, then everything is finished. If it is not empty, we repeat the above procedure to check if E_{k+2} is empty or not.

If this procedure does not end with a finite k for a certain set E_∞ of the springs , then we define that the next state of such springs is plastic. The reason is as follows. By this assignment of the next state we chose the next state for all the springs of E. The problem (1.1)~(1.2) is hence solved under this stress-strain relation. The solution then satisfies the conditions (1.2) and, especially for the springs in E_∞ , $\dot{U}_i = 0$ is satisfied after t_0. Hence this solution is just what we were seeking.

We determined a combination of the stress-strain relations just after t_0 for the springs in E_1. These relations, of course, are physically admissible. It is also evident by the above discussion that this combination is unique among 2^M possibilities. Hence, taking into account the energy estimate in the next section, the problem (1.1)~(1.2) has a unique solution in I.

ENERGY INEQUALITIES

The following theorem is proved in the same manner as in the dynamic problem with the kinematic hardening rule.

THEOREM 4.6. Assume that the multiple mass system is in stage(m). We have the following :

(1). The equilibrium equation (1.1) is expressed as follows :

$$(1.17)\qquad k_iU_i - \xi_i k_i \sum_{j=0}^{m_i} (-1)^j (U_i - U_i^{(j)}) - (k_{i+1}U_{i+1} - \xi_{i+1}k_{i+1} \sum_{j=0}^{m_{i+1}} (-1)^j (U_{i+1} - U_{i+1}^{(j)})) = b_i(t) \qquad i = 1 \sim N.$$

(2). The following equalities hold :

$$\frac{1}{2}\sum_{i=1}^{N+1} (k_iU_i^2 - \xi_i k_i \sum_{j=0}^{m_i} (-1)^j (U_i - U_i^{(j)})^2) = \sum_{i=1}^{N} \int_0^t b_i \dot{u}_i \, dt.$$

$$\sum_{i=1}^{N+1} (k_i(\dot{U}_i)^2 - \xi_i k_i \sum_{j=0}^{m_i} (-1)^j (\dot{U}_i)^2) = \sum_{i=1}^{N} \dot{b}_i \dot{u}_i.$$

(3). The following inequalities hold from which the boundedness of $\sum U_i^2$ and of $\sum \dot{U}_i^2$ follow.

$$\frac{1}{4}\sum_{i=1}^{N+1} (1 - \xi_i)k_iU_i^2 \le \sum_{i=1}^{N} [\frac{\xi_i(1+\xi_i)k_i}{2(1-\xi_i)} (U_i^{(0)})^2 + \int_0^t b_i \dot{u}_i \, dt]$$

$$\sum_{i=1}^{N+1} (1 - \xi_i)k_i\dot{U}_i^2 \le \sum_{i=1}^{N} \dot{b}_i \dot{u}_i.$$

A WEAK FORM

THEOREM 4.7. The problem (1.1)~(1.2) is equivalent to the following problem : Seek $(u, \sigma, \alpha) \in C_+^1(I)$ satisfying

$$(1.18)\qquad \sigma_i - \sigma_{i+1} = b_i(t) \qquad i = 1 \sim N$$

$$(1.19)\qquad \sum_{i=1}^{N+1} (\dot{\sigma}_i - k_i\dot{U}_i)(\tau_i - \sigma_i) \ge 0 \qquad \text{for all } \tau \in K_\alpha$$

$$(1.20)\qquad \dot{\alpha}_i = (1 - \frac{1}{\xi_i})(\dot{\sigma}_i - k_i\dot{U}_i) \qquad i = 1 \sim N+1$$

with $\sigma \in K_\alpha$, $u_i(0) = \sigma_i(0) = \sigma_{N+1} = \alpha_i(0) = 0$, where

$$K_\alpha = \{ \tau \in C^1_+(I) \ ; \ \underset{i}{\mathrm{Max}} \ |\tau_i - \alpha_i| \le z_0 \qquad \text{for all } t \in I \}.$$

4.2. Isotropic hardening problem

The quasi-static problem of the multiple mass system with the isotropic hardening assumption is to seek (u, σ, U^p) satisfying the following equations:

$$\sigma_i - \sigma_{i+1} = b_i(t) \qquad i = 1 \sim N \tag{2.1}$$

$$\begin{cases} \dot{\sigma}_i = k_i \dot{U}_i, \quad \dot{U}^p_i = 0 & \text{if } |\sigma_i| < H(\bar{U}^p_i) \text{ or } |\sigma_i| = H(\bar{U}^p_i) \text{ and } \sigma_i \dot{\sigma}_i < 0 \\ \dot{\sigma}_i = (k_i - \dfrac{k_i^2}{g(\sigma_i)+k_i})\dot{U}_i, \quad \dot{U}^p_i = \dfrac{\dot{\sigma}_i}{g(\sigma_i)} & \text{if } |\sigma_i| = H(\bar{U}^p_i) \text{ and } \sigma_i \dot{\sigma}_i \ge 0, \end{cases} \tag{2.2}$$

where

$$\bar{U}^p_i = \int_0^t |\dot{U}^p_i| \, dt, \qquad g(\sigma_i) = H'(H^{-1}(|\sigma_i|))$$

$$u(0) = \sigma(0) = U(0) = 0, \ u_0 = u_{N+1} = 0, \ b_i(0) = 0.$$

The argument applied to the case of kinematic hardening to continuate the solution is essentially valid in this problem. To make sure of it, we describe the procedure briefly.

Until $t = t_0$ at which some spring satisfies $|\sigma_i| = H(0) = z_0$, all the springs are elastic, so the problem to be solved first is

$$\sigma_i - \sigma_{i+1} = b_i(t) \qquad i = 1 \sim N$$

$$\sigma_i = k_i U_i, \quad \dot{U}^p_i = 0.$$

Let E_0 be the set of $i \in E$ for which $|\sigma_i| = H(0)$ is satisfied at $t = t_0$. We want to know the sign of $\dot{U}_i(t_0 + 0)$ for $i \in E_0$. The springs in $E - E_0$ are still elastic after t_0, of course. We seek the solution (u^0, σ^0) of the following problem set up at $t = t_0 + 0$:

(2.3) $$\sigma_i^0 - \sigma_{i+1}^0 = \dot{b}_i(t_0 + 0) \qquad i = 1 \sim N$$

(2.4) $$\sigma_i^0 = k_i U_i^0 \qquad \text{for } E - E_0$$

(2.5) $$\begin{cases} \sigma_i^0 = k_i U_i^0 & \text{in } D_- = \{ u^0 \in R^N ; \sigma_i(t_0) U_i^0 < 0 \} \\ \sigma_i^0 = (k_i - \dfrac{k_i^2}{g(\sigma_i) + k_i}) U^0 & \text{in } D_+ = \{ u^0 \in R^N ; \sigma_i(t_0) U_i^0 \geq 0 \} \end{cases}$$

for E_0,

where $U_i^0 = u_i^0 - u_{i-1}^0$ and $u_0^0 = u_{N+1}^0 = 0$.

THEOREM 4.8. There exists a unique solution (u^0, σ^0) of (2.3)~(2.5). u^0 is the minimizing point of the functional

$$F_1(u^0) = \frac{1}{2}(\sigma^0, U^0) - (\dot{b}(t_0 + 0), u^0)$$

under the conditions (2.4)~(2.5).

The proof of this theorem is almost the same as that of Theorem 4.1 except that the hyperplane X_i is defined by

$$X_i = \{ u^0 \in R^N ; \sigma_i(t_0) U^0 = 0 \}$$

in the present case. Hence we omit the proof. Now let (u^0, σ^0) be the solution of (2.3)~(2.5). According to the sign of $\sigma_i(t_0 + 0) U_i^0$ we classify E_0 as $E_0 = E_0^e + E_0^p$, where

$$E_0^e = \{ i \in E_0 ; \sigma_i(t_0) U_i^0 < 0 \}, \quad E_0^p = \{ i \in E_0 ; \sigma_i(t_0) U_i^0 \geq 0 \}$$

and solve the following problem for $t \geq t_0$:

(2.6) $$\sigma_i - \sigma_{i+1} = b_i(t) \qquad i = 1 \sim N$$

$$(2.7)\quad \begin{cases} \dot{\sigma}_i = k_i \dot{U}_i, \quad \dot{U}_i^p = 0 & \text{for } i \in E - E_0^p \\ \dot{\sigma}_i = (k_i - \dfrac{k_i^2}{g(\sigma_i)+k_i})\dot{U}_i, \quad \dot{U}_i^p = \dfrac{\dot{\sigma}_i}{g(\sigma_i)} & \text{for } i \in E_0^p \end{cases}$$

where $(u, \sigma, U^p)(t_0) = (u(t_0), \sigma(t_0), 0)$. Then it is easy to see that this problem has a unique solution (u,σ,U^p) at least in a certain neighborhood of t_0 and $(\dot{u},\dot{\sigma})(t_0+0) = (u^0,\sigma^0)$. Let $E_1 = \{ i \in E_0^p \,;\, \dot{U}_i(t_0+0) = 0\}$. The spring $i \in E - E_0^p$ and $i \in E_0^p - E_1$ must be elastic and plastic, respectively, after t_0. If E_1 is empty, then the solution is continuated beyond t_0. But even when it is not empty, $(\dot{u}, \dot{\sigma})(t_0 + 0)$ is independent of the choice of the next state of the points of E_1, as is proved in the same manner as before. Hence we can repeat the above method to find the sign of $\ddot{U}_i(t_0+0)$ for $i \in E_1$, and of the higher derivatives if $\ddot{U}_i(t_0+0)$ vanishes. The general procedure is as follows :

Let E_k $(k \geq 1)$ be a subset of E_0 such that it is empty or, if not, satisfies the followings :

For all the springs in E_k the next state is assigned at $t = t_0$(it may be tentative). Under this stress-strain relation we solve the problem (2.1) for $t \geq t_0$. Then the following holds.

(A). The quantities

$$\frac{d^r}{dt^r}(u_i,\sigma_i)(t_0+0) \qquad i \in E \quad (\text{for all } r \leq k)$$

are independent of the choice of the next state of E_k.

(B). For the spring i in $E_0 - E_k$, there is an n_i $(1 \leq n_i \leq k)$ depending on i such that

$$\frac{d^r}{dt^r}U_i(t_0+0) = 0 \quad (1 \leq r < n_i), \quad U_i^{(n_i)} \equiv \frac{d^{n_i}}{dt^{n_i}}U_i(t_0+0) \neq 0 \,,$$

and the next state is so assigned that the point is elastic (or plastic) after t_0 if $\sigma_i(t_0)U_i^{(n_i)}$ is negative (or positive). (For k=1 the first condition is unnecessary). The next state of the springs in $E - E_0$ is still elastic, of course.

(C). For all the springs in E_k it holds that

$$\frac{d^r}{dt^r} U_i(t_0 + 0) = 0 \qquad (1 \leq r \leq k).$$

The next state of E_k (it was tentative) is then determined as follows :

(Step 1). Seek the solution (u^0, σ^0) of the problem :

$$\sigma_i^0 - \sigma_{i+1}^0 = \frac{d^{k+1}}{dt^{k+1}} b_i(t+0) \qquad i = 1 \sim N$$

$$(2.8) \quad \begin{cases} \sigma_i^0 = k_i U_i^0 & \text{for the elastic spring in } E - E_k \\ \sigma_i^0 = \left(k_i - \dfrac{k_i^2}{g(\sigma_i(t_0)) + k_i}\right) U_i^0 & \text{for the plastic spring in } E - E_k \end{cases}$$

and for those in E_k

$$(2.9) \quad \begin{cases} \sigma_i^0 = k_i U_i^0 & \text{in } D_- = \{u^0 \in R^N \; ; \; \sigma_i(t_0) U_i^0 < 0\} \\ \sigma_i^0 = \left(k_i - \dfrac{k_i^2}{g(\sigma_i(t_0)) + k_i}\right) U^0 & \text{in } D_+ = \{u^0 \in R^N \; ; \; \sigma_i(t_0) U_i^0 \geq 0\} \end{cases}$$

where $U_i^0 = u_i^0 - u_{i-1}^0$ and $u_0^0 = u_{N+1}^0 = 0$.

(Step 2). Using the solution (u^0, σ^0) of the above problem , classify E_k as $E_k = E_k^e + E_k^p$, where

$$E_k^e = \{ i \in E_k \; ; \; \sigma_i(t_0) U_i^0 < 0 \}, \quad E_k^p = \{ i \in E_k \; ; \; \sigma_i(t_0) U_i^0 \geq 0 \}.$$

(Step 3). The springs in E_k^e (or E_k^p) are regarded as elastic (or plastic) after t_0.

If there are some springs in E_k^p for which $\sigma_i(t_0) U_i^0 = 0$ holds, we denote the set of all such springs by E_{k+1}. Then, since E_{k+1} satisfies the conditions (A), (B) and (C) (replacing k by k+1), we can repeat the above argument to determine the next state of the springs in E_{k+1}. By this procedure the next state of all the springs in E_0 can be determined. In fact, the following theorem, which can be proved in the same way as in the kinematic hardening case, ensures the validity of this procedure.

THEOREM 4.9. (1). The solution (u^0,σ^0) of the first step exists uniquely. u^0 is characterized as the minimum point of the functional

$$F_{k+1}(u^0) = \frac{1}{2}(\sigma^0, U^0) - \left(\frac{d^{k+1}}{dt^{k+1}} b(t_0+0), u^0\right)$$

under the conditions (2.8) and (2.9).

(2). Let (u,σ,U^p) be the solution of

$$\sigma_i - \sigma_{i+1} = b_i(t) \qquad i = 1 \sim N$$

$$\begin{cases} \dot\sigma_i = k_i\dot U_i, \quad \dot U^p_i = 0 & \text{for } E^e_k \text{ and the elastic spring in } E - E_k \\ \dot\sigma_i = \left(k_i - \dfrac{k_i^2}{g(\sigma_i)+k_i}\right)\dot U_i, \quad \dot U^p_i = \dfrac{\dot\sigma_i}{g(\sigma_i)} & \text{for } E^p_k \text{ and the plastic spring in } E - E_k \end{cases}$$

with the initial condition $(u,\sigma,U^p)(t_0) = (u,\sigma,U^p)(t_0-0)$. Then it holds that

$$\frac{d^{k+1}}{dt^{k+1}}(u,\sigma)(t_0+0) = (u^0,\sigma^0).$$

(3). The value of the left side of the above equality is determined independently of the choice of the next state of E_{k+1}.

ENERGY INEQUALITIES AND THE WEAK FORM

The following theorems are now easily verified.

THEOREM 4.10. Let (u,σ,U^p) be the solution of (2.1)~(2.2). Then it holds that

$$\sum_{i=1}^{N+1} \left(\frac{1}{2k_i}\sigma_i^2 + \int_0^t \sigma_i \dot{U}_i^p \, dt\right) = \sum_{i=1}^{N} \int_0^t b_i \dot{u}_i \, dt$$

$$\sum_{i=1}^{N+1} \frac{1}{k_i} \dot{\sigma}_i^2 \leq \sum_{i=1}^{N} \dot{b}_i \dot{u}_i .$$

THEOREM 4.11. The problem (2.1)~(2.2) is equivalent to the following : Seek $(u,\sigma,\hat{U}^p) \in C_+^1(I)$ satisfying

$$\sigma_i - \sigma_{i+1} = b_i \qquad i = 1 \sim N$$

$$\sum_{i=1}^{N+1} [(\dot{U}_i - \frac{1}{k_i}\dot{\sigma}_i)(\tau_i - \sigma_i) - \dot{\hat{U}}_i^p(\zeta_i - \hat{U}_i^p)] \leq 0 \qquad \text{for all } (\tau_i, \zeta_i) \in K_G$$

with $(\sigma_i, \hat{U}_i^p) \in K_G$, $u_i(0) = \sigma_i(0) = \hat{U}_i^p(0) = 0$, where $\hat{U}_i^p$ and K_G are those defined before.

CHAPTER 5

TWO-DIMENSIONAL DYNAMIC SEMIDISCRETE SYSTEM

5.1. A finite element approximation

In this chapter we consider a semidiscrete finite element approximation of the two-dimensional dynamic problem with kinematic hardening rule. It was formulated in Chapter 1 as follows :

$$(1.1) \qquad \rho\ddot{u}_i - \sum_{j=1}^{2} \sigma_{ij,j} = b_i \qquad \text{in } I\times\Omega \quad (i = 1,2)$$

$$(1.2) \qquad \begin{cases} \dot{\sigma} = D\dot{\varepsilon} \\ \dot{\alpha} = 0 \end{cases} \qquad \text{if } f(\sigma - \alpha) < z_0$$

$$(1.3) \qquad \begin{cases} \dot{\sigma} = (D - D')\dot{\varepsilon} \\ \dot{\sigma} = (\sigma - \alpha)\,\dfrac{\partial f*\dot{\sigma}}{f} \end{cases} \qquad \text{if } f(\sigma - \alpha) = z_0 \text{ and } \partial f*\dot{\sigma} \geq 0$$

The yield surface is given by $f(\sigma - \alpha) = z_0$ for a certain positive constant z_0.

We shall introduce a typical finite element approximation of this problem. We assume that Ω is a polygonal region and denote its triangulation by the same notation Ω. We require that the end points of Γ_u on which the zero boundary condition be given are the nodes of Ω. Now let $\{\phi_p\}$ be the usual piecewise linear finite element basis which takes 1 at the node p. The approximate value of u_i at time t is sought in the form

$$(1.4) \qquad u_i(t) = \sum_{p\in P} u_i^p(t)\phi_p,$$

where P is the set of nodes in $\Omega - \Gamma_u$. The unknowns $\{u_i^p(t)\}$ are determined by the following Galerkin system :

$$(1.5)\qquad (\rho\ddot{u}_i,\phi_p) + \sum_j (\sigma_{ij},\phi_{p,j}) = (b_i,\phi_p) \qquad p \in P,$$

$$(1.6)\qquad \begin{cases} \dot{\sigma} = D\dot{\varepsilon} \\ \dot{\alpha} = 0, \end{cases} \qquad \text{for elastic elements}$$

$$(1.7)\qquad \begin{cases} \dot{\sigma} = (D - D')\dot{\varepsilon} \\ \dot{\alpha} = (\sigma - \alpha)\,\dfrac{\partial f * \dot{\sigma}}{f} \end{cases} \qquad \text{for plastic elements}$$

The displacements u_i are linear on each triangle (element), so the strains ε_{ij} are constant on it. Hence σ and α which must be determined by (1.6) or (1.7) are also constant on each element. Therefore, we define an element as elastic (or plastic) if

$$f(\sigma - \alpha) < z_0 \qquad (\text{ or } f(\sigma - \alpha) = z_0 \text{ and } \partial f * \dot{\sigma} \geq 0)$$

is satisfied on that element. The strain-displacement relation and the initial conditions are the same as in the continuous problem. We assume that the initial velocity is interpolated by $\{\phi_p\}$. We call (1.5)~(1.7) the semi-discrete system since the spatial region is discretized but time is still continuous.

REMARK. If we choose suitable basis functions for a rod problem, then the semidiscrete system reduces to the equations of the multiple spring - mass system which is treated in Chapter 3. Assume a straight uniform rod submitted to a longitudinal impact. The simplest mathematical model to represent the elastic-plastic vibration of this rod is given by the following differential equations :

(1.8) $$\rho\ddot{u} - \sigma_x = b \qquad \text{in } I\times\Omega$$

where

(1.9) $$\dot{\sigma} = \begin{cases} k\dot{u}_x & \text{in the elastic region} \\ (1-\xi)k\dot{u}_x & \text{in the plastic region,} \end{cases}$$

and ξ, k are positive constants ($\xi < 1$). Let Ω be the interval (0,1), and consider the equation (1.8) in $I\times\Omega$. We divide Ω into N elements of equal length h. Let i be the point ih (i = 0, 1,..., N) and e_i the element [(i-1)h,ih]. We use different basis functions :

$\phi_i(x)$: usual piecewise linear basis

$\bar{\phi}_i(x)$: characteristic function of $[ih - \frac{h}{2}, ih + \frac{h}{2}]$.

Consider the following system of equations :

(1.10) $$\rho(\ddot{\bar{u}},\bar{\phi}_i) + \sum_e (\sigma(u),\phi_{i,x})_e = (b,\phi_i) \qquad i = 1,\dots, N$$

(1.11) $$\dot{\sigma}(u) = \begin{cases} k\dot{u}_x & \text{in the elastic region} \\ (1-\xi)k\dot{u}_x & \text{in the plastic region,} \end{cases}$$

where

$$u = \sum_{i=1}^{N} u_i(t)\phi_i(x) , \qquad \bar{u} = \sum_{i=1}^{N} u_i(t)\bar{\phi}_i(x).$$

Now since σ is constant on each element, we put

$$\sigma_i = \sigma(u)|_{e_i}$$

to get

$$\sum_e (\sigma(u),\phi_{i,x})_e = \sigma_i - \sigma_{i+1} \qquad (\sigma_{N+1} = 0).$$

Then the system (1.10)~(1.11) is written as

$$\text{(1.12)} \qquad \rho_i \ddot{u}_i + \sigma_i - \sigma_{i+1} = b_i \qquad i = 1, \ldots, N$$

$$\text{(1.13)} \qquad \dot{\sigma}_i = \begin{cases} \frac{k}{h} (\dot{u}_i - \dot{u}_{i-1}) & \text{for the elastic element} \\ (1 - \xi) \frac{k}{h} (\dot{u}_h - \dot{u}_{i-1}) & \text{for the plastic element} \end{cases}$$

which is the very same system considered in the previous chapter. Here $\rho_i = \rho h$ for $i \neq N$ and $\rho_i = \rho h/2$ for $i = N$, and $u_0 = 0$.

5.2. Continuation of the solution

The semidiscrete system is a starting point from which to derive numerical methods for the present problem , and it can also be used to get the solution of the fully continuous problem. Hence we first seek the solution (u,σ,α) of this system. The key is again the continuation of the solution beyond the time at which the yielding or unloading may occur. Anyway, it is evident that all the elements are elastic until the condition $f(\sigma) = z_0$ is satisfied by some element. Suppose that this condition is satisfied at $t = t_0$ by the elements of $\bar{E}_0$ which is a subset of E, the set of all the elements. The elements of $E - \bar{E}_0$ remain elastic after t_0. We want to prove

THEOREM 5.1. Each element of $\bar{E}_0$ is either elastic or plastic after t_0. This determination depends on the data of the solution before t_0 plus the data of b.

The proof of this theorem consists of three lemmas. Take an element of $\bar{E}_0$ and ε, σ, α on this element. As is proved in the final step of the following discussion, we can assume that ε, σ and α are analytic, at least in a certain time interval $[t_0, t_0+\delta)$ ($\delta>0$).

LEMMA 5.2. Suppose that the sign of $\partial f^*_{(\sigma)}\dot{\sigma}|_{t_0+0}$ for the elements of $\bar{E}_0$ is determined independently of the choice of the next σ - ε relation of the element of $\bar{E}_0$. Then there is a positive constant δ such that the following hold in $I_\delta = (t_0, t_0+\delta)$.

(1). For the element of $\bar{E}_0$ satisfying $\partial f^*_{(\sigma)}\dot{\sigma}|_{t_0+0} < 0$, set $\dot{\alpha} = 0$ for $t > t_0$. Then for any choice of the next state of $\bar{E}_0$ it holds that

$$f(\sigma) < z_0 \qquad \text{in } I_\delta.$$

(2). For the element of $\bar{E}_0$ satisfying $\partial f^*_{(\sigma)}\dot{\sigma}|_{t_0+0} > 0$, set $\dot{\alpha} = (\sigma-\alpha)\,\partial f^*_{(\sigma-\alpha)}\dot{\sigma}/f(\sigma-\alpha)$ for $t > t_0$. Then for any choice of the next state of $\bar{E}_0$ it holds that

$$f(\sigma - \alpha) = z_0, \qquad \partial f^*_{(\sigma-\alpha)}\dot{\sigma} > 0 \qquad \text{in } I_\delta.$$

PROOF. If $\partial f^*\dot{\sigma}$ is negative at $t = t_0+0$, then by the continuity of $\dot{\sigma}$ with respect to $t \geq t_0+0$ we have

$$f^2(\sigma) - z_0^2 = \int_{t_0}^{t} \frac{d}{d\tau} f^2(\sigma(\tau))\,d\tau = 2\int_{t_0}^{t} f\,\partial f^*_{(\sigma)}\dot{\sigma}\,d\tau < 0 \qquad (t \in I_\delta)$$

which proves (1). In the same way (2) is also proved since if $\dot{\alpha}$ satisfies the above equation, then it holds that

$$\frac{d}{dt} f(\sigma - \alpha) = \partial f^*_{(\sigma-\alpha)}(\dot{\sigma} - \dot{\alpha}) = 0 \; ;$$

that is, $f(\sigma - \alpha) = \text{const.}$ after t_0.

By this lemma we see that if the sign of $\partial f^*\dot{\sigma}|_{t_0+0}$ is negative (or positive) then the next state of this element must be elastic (or plastic).

We here emphasize that the sign of this quantity is determined only by the data before t_0. To see this, suppose that we solve the problem by arbitrarily assuming the next state of the elements of $\bar{E}_0$. We then have its solution (u,σ,α) uniquely, at least in a certain neighborhood of t_0. Next we check the sign of $\partial f^*\dot{\sigma}|_{t_0+0}$ for the element with which we are now concerned. If the next state is assumed to be elastic, then this quantity is equal to $\partial f^*D\dot{\varepsilon}|_{t_0+0}$. If it is plastic, then it is equal to

$$\partial f^*(D\dot{\varepsilon} - \frac{D\partial f\partial f^*D}{\eta+\partial f^*D\partial f}\dot{\varepsilon})|_{t_0+0}$$

$$= \partial f^*D\dot{\varepsilon}|_{t_0+0}(1 - \theta) \qquad (0 < \theta < 1).$$

Since $\dot{\varepsilon} = \varepsilon(\dot{u})$ is continuous in the dynamic problem, the sign of $\partial f^*\dot{\sigma}$ itself is the same in either case.

For the elements of $\bar{E}_0$ satisfying $\partial f^*\dot{\sigma}|_{t_0+0} \neq 0$, therefore, only one choice of the next state is admissible. Thus we suppose that the next state is already chosen for this kind of element, and we exclude them from consideration.

Let E_0 be the set of elements of $\bar{E}_0$ for which $\partial f^*\dot{\sigma}|_{t_0+0} = 0$ holds. If E_0 is empty, the proof of Theorem 5.1 is complete. We shall discuss below the case where E_0 is not empty and the next state of the elements of E_0 still cannot be chosen. In this case we have to examine the higher derivative of $\partial f^*\dot{\sigma}$ at $t = t_0+0$. The next lemma is a general version of Lemma 5.2 and is readily proved. Suppose that we solve the problemm by arbitrarily assuming the next state of E_0. In what follows ∂f denotes $\partial f(\sigma-\alpha)$.

LEMMA 5.3. Assume that σ and α for an element satisfy $f(\sigma - \alpha) = z_0$ and $\partial f^*\dot{\sigma} = 0$ at $t = t_0 + 0$. Expand $\partial f^*\dot{\sigma}$ in the Taylor series :

$$g(t) = \partial f^*\dot{\sigma} = \sum_k g_0^{(k)}(t - t_0)^k.$$

Let k_0 ($< \infty$) be such an integer (if it exists) that

$$g_0^{(k_0)} \neq 0, \quad g_0^{(k)} = 0 \qquad (k < k_0).$$

Then there is a positive constant δ such that in $I_\delta = (t_0, t_0 + \delta)$ it holds that

(1) if $g_0^{(k_0)} < 0$ and $\dot{\alpha} = 0$ for $t > t_0$, then

$$f(\sigma - \alpha) < z_0 \quad \text{in } I_\delta,$$

(2) if $g_0^{(k_0)} > 0$ and $\dot{\alpha} = (\sigma - \alpha)\partial f^* \dot{\sigma} / f(\sigma - \alpha)$ for $t > t_0$, then

$$\partial f^* \dot{\sigma} > 0, \quad f(\sigma - \alpha) = z_0 \quad \text{in } I_\delta.$$

It is clear from this lemma that if $g_0^{(k_0)}$ is negative (or positive) then the next state of this element must be elastic (or plastic). In what follows we shall show that the sign of this quantity can be determined in advance by using the data before t_0.

(I). We first note that for the elements of E_0 it holds that

(a) $\partial f^* D\dot{\varepsilon}|_{t_0+0} = 0$,

(b) $\dot{\sigma}|_{t_0+0}$ is independent of the next state of the element of E_0.

The first assertion is evident. The second follows from the continuity of $\dot{\varepsilon}$.

(II). Let E_k ($k \geq 0$) be the set of elements for which $f(\sigma - \alpha) = z_0$ holds and the sign of the derivative of order i ($i \leq k$) of the function $\partial f^* \dot{\sigma}$; that is, the sign of $(\partial f^* \dot{\sigma})^{(i)}$ ($i \leq k$) vanishes at $t = t_0 + 0$ independently of the next state of E_k. The next state of $E - E_k$ is assumed to be already determined independently of the choice of the next state of E_k. Then we have

LEMMA 5.4. Suppose that for all the elements of E_k it holds that

(A) $(\partial f^* D\dot{\varepsilon})_e^{(i)}|_{t_0+0} = 0 \qquad (i \le k)$,

(B) $\sigma_e^{(i+1)}|_{t_0+0}$ $(i \le k)$ are determined independently of the next state of E_k,

(C) $u^{(i+2)}|_{t_0+0}$ $(i \le k)$ are determined independently of the next state of E_k.

We solve the problem by arbitrarily assuming the next state of E_k. We then have

(α) For all the elements of E_k, the sign of $(\partial f^*\dot{\sigma})^{(k+1)}|_{t_0+0}$ is independent of the next state of E_k. Hence the next state can be assigned to the elements of E_k for which this quantity is not zero, and

(β) For the elements of E_k for which $(\partial f^*\dot{\sigma})^{(k+1)}|_{t_0+0} = 0$ holds (note that the totality of the elements satisfying this condition is E_{k+1} by definition) the above (A), (B) and (C) hold, replacing k by k + 1.

PROOF. Since (A) holds, there are two possibilities according to the next state :

$$(\partial f^*\dot{\sigma})^{(k+1)}|_{t_0+0} = \begin{cases} (\partial f^* D\dot{\varepsilon})^{(k+1)}|_{t_0+0} \\ (\partial f^* D\dot{\varepsilon})^{(k+1)}|_{t_0+0}\,(1-\theta). \end{cases}$$

By (B) and (C), $(\partial f^* D\dot{\varepsilon})^{(k+1)}$ is independent of the next state of E_k. This proves (α). If $(\partial f^*\dot{\sigma})^{(k+1)} = 0$ at $t_0 + 0$, then so is $(\partial f^* D\dot{\varepsilon})^{(k+1)}$ by this result. Hence (A) follows, replacing k by k+1. To prove (B) for k+1, we note that either

$$\sigma^{(k+2)} = D\varepsilon^{(k+2)},$$

or

(2.1) $$\sigma^{(k+2)} = D\varepsilon^{(k+2)} - \frac{1}{\eta} D \left(\sum_{r=0}^{k+1} {}_{k+1}C_r (\partial f)^{(k+1)} (\partial f^* \dot{\sigma})^{(r)} \right)$$

holds after t_0. This follows from the relation

(2.2) $$\dot{\sigma} = D\dot{\varepsilon} - \frac{1}{\eta} D \partial f \partial f^* \dot{\sigma},$$

which is a modification of the plastic $\sigma - \varepsilon$ relation. Now the second term of the right side of (2.1) vanishes at $t = t_0 + 0$. Hence (B) holds for k+1 by (C) of the assumption. Finally, we have

(2.3) $$(\rho u_i^{(k+3)}, \phi_p) + \sum_j (\sigma_{ij}^{(k+1)}, \phi_{p,j}) = (b_i^{(k+1)}, \phi_p) \qquad p \in P$$

for $t \geq t_0 + 0$. The next state of the elements of $E - E_k$ is already known and thus $\sigma^{(i)}|_{t_0+0}$ $(i \leq k+1)$ for such elements is independent of the next state of E_k, since these values are uniquely determined by the derivatives of u of order $\leq$ k+1, and these derivatives are independent of the next state of E_{k+1} by (C) of the assumption. On the other hand, the quantity $\sigma^{(k+1)}|_{t_0+0}$ of the element of E_k is, of course, independent of the next state of E_{k+1}, so that (2.3) implies that $u^{(k+3)}|_{t_0+0}$ is determined independently of the next state of E_{k+1}. Hence (C) holds well.

PROOF OF THEOREM 5.1. For the elements of E_0, the conditions (A),(B) and (C) hold as we have seen in (I) (the condition (C) follows from the continuity of $u^{(2)}$). The set E_1 is thus defined, and the conditions (A),(B) and (C) hold . The next state may be chosen for some elements of E_1 as we desired. This procedure can be continued until the next state is determined for all the elements of E_0 , except for the special elements which satisfy $(\partial f^* \dot{\sigma})^{(k)}|_{t_0+0} = 0$ for all k. In this case, we define that these elements are plastic after t_0 . The reason is exactly the same as in the spring - mass system. Note that this case corresponds to the neutral state, and the stress point σ moves (or settles) on a fixed yield surface.

REMARK. So far we have discussed only the initial yielding. The solution can be continuated beyond that time. However, Theorem 5.1 itself is valid for the subsequent yieldings and also for the case of unloading, as was readily seen in the above discussion. Hence the initial value problem of this semi-discrete system can be set up at any time.

5.3. Energy inequalities

To continuate the solution over the whole time interval we need to know the boundedness of the solution. In this section we derive two a priori estimates which are independent of the triangulation of the domain.

Let (u, σ, α) be the solution of the semidiscrete system (in a time interval in which its existence is assured). Introduce $E_0(t)$ and $E_1(t\pm 0)$ by

$$E_0(t) = \|\dot{u}\|_\rho^2 + \frac{1}{\eta}\|\alpha\|_S^2 + \|\sigma\|_C^2,$$

$$E_1(t\pm 0) = (\|\ddot{u}\|_\rho^2 + \frac{1}{\eta}\|\dot{\alpha}\|_S^2 + \|\dot{\sigma}\|_C^2)(t\pm 0),$$

where $\|u\|_\rho^2$ and $\|\sigma\|_A^2$ (A: matrix) denote $\sum(\rho u_i, u_i)$ and $(A\sigma,\sigma)$, respectively.

THEOREM 5.5. $E_0(t)$ is bounded uniformly on the triangulation of Ω.

PROOF. It is easy to see that the vector ∂f and $\sigma - \alpha$ are connected by the relation

$$f\partial f = S(\sigma - \alpha), \qquad 2S = \begin{pmatrix} 2 & -1 & 0 \\ -1 & 2 & 0 \\ 0 & 0 & 6 \end{pmatrix}. \tag{3.1}$$

Therefore, in any state we have the relation

(3.2) $$S\dot{\alpha} = \eta(\dot{\varepsilon} - C\dot{\sigma}).$$

Multiplying both sides of this equality by σ, we have by (1.5)

$$0 = -(\sigma,\dot{\varepsilon}) + \frac{1}{\eta}(\sigma,S\dot{\alpha}) + (\sigma,C\dot{\sigma})$$

$$= \frac{1}{2}\frac{d}{dt}\left(\|\dot{u}\|_\rho^2 + \frac{1}{\eta}\|\alpha\|_S^2 + \|\sigma\|_C^2 \right) - (b,\dot{u}) + \frac{1}{\eta}(\sigma - \alpha, S\dot{\alpha}).$$

The last term is non - negative by (2.2), (3.1) and (3.2). Since $(\dot{u},\sigma,\alpha)$ is continuous, we can integrate the last identity to get

$$E_0(t) \leq C_1 + C_2\int_0^t \|b(s)\| E_0(s)^{\frac{1}{2}}\, ds$$

to which lemma 2.7 is applicable. This completes the proof.

To derive the estimate of the higher derivatives, we prepare

LEMMA 5.6. Assume that the stress point σ leaves the yield surface $\{\tau \in R^3 ; f^2(\tau-\alpha) = z_0^2\}$ at time $t = t_0$. Then it holds that

$$\lim_{t \to t_0 \pm 0} \partial f^* \dot{\sigma} = 0.$$

REMARK. This is why we exclude the condition " $f(\sigma-\alpha) = z_0$ and $\partial f^*\dot{\sigma} < 0$ " from the elasticity condition which becomes necessary for quasi - static problems.

PROOF OF THE LEMMA. There are three cases to be checked.

(1). In a certain neighborhood of t_0, t_0 is the only t which satisfies the equality $f(\sigma - \alpha) = z_0$. In this case we have

$$0 > \partial f^*|_{t_0} (\sigma(t_0+\delta) - \sigma(t_0))$$

for small $\delta > 0$. Hence we have

(3.3) $$0 \geq \partial f^*\dot{\sigma}|_{t_0+0} = \partial f^*D\dot{\varepsilon}|_{t_0+0}$$

(3.4) $$0 \leq \partial f^*\dot{\sigma}|_{t_0-0} = \partial f^*D\dot{\varepsilon}|_{t_0-0}$$

so that $\partial f^*\dot{\sigma}$ has to vanish at $t = t_0$ by the continuity of $\dot{\varepsilon}$.

(2). The element is plastic in $(t_0-\delta, t_0)$ and elastic in $(t_0, t_0+\delta)$ $(\delta > 0)$. In this case, we remember that $\partial f^*\dot{\sigma} \geq 0$ holds in plastic state. Suppose that

(3.5) $$\partial f^*\dot{\sigma}|_{t_0-0} > 0.$$

Since we have in the plastic state

$$\partial f^*\dot{\sigma} = \partial f^*D\dot{\varepsilon}\left(1 - \frac{\partial f^*D\partial f}{\eta+\partial f^*D\partial f}\right),$$

(3.5) implies $\partial f^*D\dot{\varepsilon} > 0$ at $t = t_0 - 0$. On the other hand, as is seen from (3.3), $\partial f^*D\dot{\varepsilon}$ must be non-positive at $t = t_0 + 0$. This contradicts the fact that $\dot{\varepsilon}$ is continuous. Hence $\partial f^*\dot{\sigma}|_{t_0-0} = 0$, and thus $\partial f^*\dot{\sigma}|_{t_0+0} = 0$.

(3). The case where t_0 is an accumulation point of the t's at which the state change occurs for some elements. We first note that these elements are elastic for $t \in (t_0, t_0+\delta)$ for some $\delta > 0$, since σ leaves the yield surface at t_0. Hence $\partial f^*\dot{\sigma}|_{t_0+0} = \partial f^*D\dot{\varepsilon}|_{t_0+0}$ holds. Now, by the asssumption, the unloading points accumulate to t_0. Therefore $\partial f^*D\dot{\varepsilon} \to 0$ as $t \to t_0 - 0$, which implies $\partial f^*\dot{\sigma} \to 0$ as $t \to t_0 - 0$; that is, $\partial f^*\dot{\sigma}|_{t_0-0} = \partial f^*D\dot{\varepsilon}|_{t_0-0} = 0$, and thus $\partial f^*\dot{\sigma}|_{t_0+0} = 0$ by the continuity of $\dot{\varepsilon}$. This completes the proof.

THEOREM 5.7. $E_1(t \pm 0)$ is bounded uniformly on the triangulation of Ω.

PROOF. Consider an interval $I_i = (t_0, t_1)$ on which no change of state occurs for any element. On I_i we have

$$S\ddot{\alpha} = \eta(\ddot{\varepsilon} - C\ddot{\sigma}).$$

Hence, in the same way as before, we obtain

$$(3.6) \qquad \frac{1}{2}\frac{d}{dt}\left(\|\ddot{u}\|_\rho^2 + \frac{1}{\eta}\|\dot{\alpha}\|_S^2 + \|\dot{\sigma}\|_C^2 \right) - (\dot{b}, \ddot{u}) + \frac{1}{\eta}(\dot{\sigma} - \dot{\alpha}, S\ddot{\alpha}) = 0.$$

To show that the last term of the right side is non-negative, we note that for each element it holds that either $\ddot{\alpha} = 0$ or

$$S\ddot{\alpha} = \left(\frac{d}{dt}\,\partial f\right)\partial f^*\dot{\sigma} + \partial f\frac{d}{dt}(\partial f^*\dot{\sigma}).$$

Here we have the identity

$$\frac{d}{dt}\,\partial f = \frac{1}{z_0}S(\dot{\sigma} - \dot{\alpha}).$$

Since $\partial f^*(\dot{\sigma} - \dot{\alpha}) = 0$ in the plastic state, we have

$$(\dot{\sigma} - \dot{\alpha})^*S\ddot{\alpha} = \frac{1}{z_0}(\dot{\sigma} - \dot{\alpha})^*S(\dot{\sigma} - \dot{\alpha})\ \partial f^*\dot{\sigma} \geq 0.$$

We now integrate (3.6) in I_i to get

$$(3.7) \qquad E_1(t_0 + 0) + 2\int_{t_0}^{t_1}(\dot{b}, \ddot{u})\,dt \geq E_1(t_1 - 0).$$

In what follows we shall show that $E_1(t_0 - 0) \geq E_1(t_0 + 0)$ for any $t_0 \in I$.

Let e be an element and $E_1^e(t)$ the e-part of $E_1(t)$: $E_1(t) = \sum_e E_1^e(t)$.

(1). Assume that t_0 is not the accumulation point of such t's that the state changes at t for some element. Let $I_{i-1} = (t_{-1}, t_0)$ be a time interval on which no change of state occurs. We have to examine the following two cases :

(a). The case where e is elastic on I_{i-1} and plastic on I_i. On $I_i \times e$ it holds that

$$\|\dot{\alpha}\|_S^2 = (\partial f, S^{-1}\partial f)(\partial f*\dot{\sigma})^2 = (\partial f*\dot{\sigma})^2 \text{ area}(e),$$

$$(C\dot{\sigma},\dot{\sigma}) = (\dot{\varepsilon} - \frac{1}{\eta} S\dot{\alpha},\dot{\sigma}) = (\dot{\varepsilon},\dot{\sigma}) - \frac{1}{\eta}(S\dot{\alpha},\dot{\sigma})$$

$$= (\dot{\varepsilon},(D - D')\dot{\varepsilon}) - \frac{1}{\eta}(\partial f*\dot{\sigma})^2 \text{ area}(e),$$

so that on the same region it holds that

$$\frac{1}{\eta}\|\dot{\alpha}\|_S^2 + \|\dot{\sigma}\|_C^2 = (\dot{\varepsilon},(D - D')\dot{\varepsilon}).$$

Since $u \in C^2(I)$, we have

$$E_1^e(t_0+0) = \|\ddot{u}\|_\rho^2(t_0+0) + (\dot{\varepsilon},(D - D')\dot{\varepsilon})(t_0+0)$$

$$= \|\ddot{u}\|_\rho^2(t_0-0) + (\dot{\varepsilon},(D - D')\dot{\varepsilon})(t_0-0)$$

$$\leq \|\ddot{u}\|_\rho^2(t_0-0) + (\dot{\varepsilon},D\dot{\varepsilon})(t_0-0) = E^e(t_0-0).$$

(b). The case where e is plastic on I_{i-1} and elastic on I_i. Since t_0 is an unloading point, we have $\partial f*\dot{\sigma} = 0$ at $t = t_0 \pm 0$ by Lemma 5.6. Hence on e, $\|\dot{\alpha}\|(t_0-0) = 0$ and $\dot{\sigma}(t_0-0) = D\dot{\varepsilon}(t_0-0) = D\dot{\varepsilon}(t_0+0) = \dot{\sigma}(t_0+0)$, so that

$$E_1^e(t_0+0) = [\|\ddot{u}\|_\rho^2 + \|\dot{\sigma}\|_C^2](t_0+0)$$

$$= [\|\ddot{u}\|_\rho^2 + \frac{1}{\eta}\|\dot{\alpha}\|_S^2 + \|\dot{\sigma}\|_C^2](t_0-0)$$

$$= E_1^e(t_0-0).$$

(2). The case where t_0 is an accumulation point. In this case the state for $t < t_0$ is not known, but for $t > t_0$ it is either elastic or plastic. Since the continuity of $\dot{\varepsilon}$ is always assured, $\partial f*\dot{\sigma} = \partial f*D\dot{\varepsilon} = 0$ hold at $t = t_0 \pm 0$. Hence we have

$$E_1^e(t_0-0) = [\|\ddot{u}\|_\rho^2 + (\dot{\varepsilon},D\dot{\varepsilon})](t_0-0).$$

If the element is elastic for $t > t_0$, it is clear that $E_1^e(t_0+0) = E_1^e(t_0-0)$. On the other hand if it is plastic then

$$E_1^e(t_0+0) = [\|\ddot{u}\|_\rho^2 + (C\dot{\sigma},\dot{\sigma})](t_0+0)$$

$$= [\|\ddot{u}\|_\rho^2 + (\dot{\varepsilon},D\dot{\varepsilon})](t_0+0)$$

$$= E_1^e(t_0-0).$$

These estimates imply that the function $E_1^e(t)$ is non-increasing at $t = t_0$ in any case. Hence we have the inequality

$$E_1(t_0-0) + 2\int_{t_0}^{t_1}(\dot{b},\ddot{u})\,dt \geq E_1(t_1-0),$$

and continuating these estimates we obtain for any $t \in I$

$$E_1(t+0) \leq E_1(t-0) \leq E_1(0) + 2\int_0^t(\dot{b},\ddot{u})\,dt,$$

from which the boundedness of $E_1(t \pm 0)$ follows.

5.4. A weak form

The fundamental assumption of our plasticity theory is that, as is seen in (3.3) of Chapter 1, the plastic strain increment $\dot{\varepsilon}^p$ is pararell to the vector ∂f at the stress point σ, or in other words, it is orthogonal to the yield surface. This property may be used to formulate the present problem in a weak form as we had already done for the case of the spring-mass system.

Let χ^e be the characteristic function of the element e. As before we define $K = K_\alpha$ ($\alpha \in C_+^1(I)$) by

$$K = \{\tau = \sum_e \tau^e(t)\chi^e \;;\tau^e \in C_+^1(I),\;\; f(\tau-\alpha) \leq z_0\}.$$

THEOREM 5.8. Let u and $(\varepsilon,\sigma,\alpha)$ be of the form

$$u(t) = \sum_{p\in P} u^p(t)\phi_p, \qquad (\varepsilon,\sigma,\alpha)(t) = \sum_e (\varepsilon^e,\sigma^e,\alpha^e)(t)\chi^e,$$

where ε^e, etc., denote the restriction of the function on e. The initial value problem of the semidiscrete system is equivalent to the following problem : Seek $(u,\sigma,\alpha) \in C^1_+(I)$ which satisfies for all $t \in I$

$$(4.1) \qquad (\rho\ddot{u}_i,\phi_p) + \sum_j (\sigma_{ij},\phi_{p,j}) = (b_i,\phi_p) \qquad \text{for all } p \in P,$$

$$(4.2) \qquad (\dot{\varepsilon} - C\dot{\sigma},\tau - \alpha) \leq 0 \qquad \text{for all } \tau \in K,$$

$$(4.3) \qquad \dot{\alpha} = \eta S^{-1}(\dot{\varepsilon} - C\dot{\sigma}),$$

and $\sigma \in K$ with the same $u - \varepsilon$ relation and the initial conditions.

PROOF. Inequality (4.2) is nothing but the normality condition of the plastic strain increments. Also, (4.3) is already proved. Hence it is only necessary to prove the uniqueness of the solution of the above problem. Substitution of (4.3) into (4.2) yields $(S\dot{\alpha},\tau - \sigma) \leq 0$. Here τ is written $\tau = \alpha + z_0\theta$ for a suitable choice of θ satisfying $f^2(\theta) \leq 1$. Let (u',σ',α') be another solution. We have two inequalities

$$(S\dot{\alpha},\alpha + z_0\theta - \sigma) \leq 0, \qquad (S\dot{\alpha}',\alpha' + z_0\theta - \sigma') \leq 0.$$

Replace θ by $(\sigma' - \alpha')/z_0$ and $(\sigma - \alpha)/z_0$ in the first and second, respectively. We then have

$$(S(\dot{\alpha} - \dot{\alpha}'),\alpha - \alpha' - [\sigma - \sigma']) \leq 0.$$

Hence by (4.1)

$$\|\dot{u} - \dot{u}'\|_\rho^2 + \|\sigma - \sigma'\|_C^2 + \frac{1}{\eta}\|\alpha - \alpha'\|_S^2 \leq 0,$$

from which follows the uniqueness.

REMARK. The right and left derivatives of $\sigma(t)$ exist for all $t \in I$. In fact, this is clear for all t except the accumulation points for which we discussed before. Even for the accumulation points, the right derivative exists, as we proved. The existence of the left derivative is shown as follows : $\dot{\sigma}$ has two expressions,

$$\dot{\sigma} = D\dot{\varepsilon} \quad \text{and } \dot{\sigma} = \left(D - \frac{D\,\partial f \partial f^* D}{\eta + \partial f^* D \partial f}\right)\dot{\varepsilon}$$

depending on the state. If t_0 is an accumulation point of $\{t_m\}$ in the above sense, then $\partial f^* D\dot{\varepsilon}(t_m)$ ($m = 1,2,\ldots$) must converge to zero as $t_m \to t_0$ by the continuity of $\dot{\varepsilon}$ and by the fact that the unloading points converge to t_0. Hence for a given $\xi > 0$ there is $\zeta > 0$ such that $|\partial f^* D\dot{\varepsilon}(t)| < \xi$ if $0 < t_0 - t < \zeta$. Now we have for $\Delta t > 0$

$$\frac{\sigma(t_0) - \sigma(t_0 - \Delta t)}{\Delta t} = \frac{1}{\Delta t}\int_{t_0-\Delta t}^{t_0} \dot{\sigma}(t)\, dt$$

$$= \frac{D\varepsilon(t_0) - D\varepsilon(t_0 - \Delta t)}{\Delta t} + \frac{1}{\Delta t}\int_{t_0-\Delta t}^{t_0} [\dot{\sigma}(t) - D\dot{\varepsilon}(t)]\, dt$$

Therefore there is a constant C which is independent of Δt such that

$$\left| \frac{\sigma(t_0) - \sigma(t_0 - \Delta t)}{\Delta t} - \frac{D\varepsilon(t_0) - D\varepsilon(t_0 - \Delta t)}{\Delta t} \right| \leq C\xi .$$

Hence the left derivative of σ exists at t_0 , and it is equalt to $D\dot{\varepsilon}(t_0)$. Notice that α also has both the right and left derivatives for all $t \in I$, since α is determined by the identity (4.3).

5.5. Isotropic hardening problem

The dynamic problem with isotropic hardening rule was formulated in Chapter 1 as follows :

$$\rho\ddot{u}_i - \sum_j \sigma_{ij,j} = b_i \qquad \text{in } I\times\Omega \tag{5.1}$$

$$\dot{\sigma} = D\dot{\varepsilon}\ ,\ \dot{\varepsilon}^p = 0 \qquad \text{if } f(\sigma) < H(\bar{\varepsilon}^p), \tag{5.2}$$

$$\dot{\sigma} = \Big(D - \frac{D\partial f\partial f^* D}{H'+\partial f^* D\partial f}\Big)\dot{\varepsilon}\ ,\ \dot{\varepsilon}^p = \frac{1}{H'}\partial f\partial f^*\dot{\sigma} \quad \text{if } f(\sigma) = H(\bar{\varepsilon}^p) \text{ and } \partial f^*\dot{\sigma} \geq 0, \tag{5.3}$$

where

$$\bar{\varepsilon}^p = \int_0^t \frac{\sigma^*\dot{\varepsilon}^p}{f(\sigma)}\,dt.$$

We consider the piecewise linear finite element approximation of this problem:

$$\rho(\ddot{u}_i,\phi_p) + \sum_j (\sigma_{ij},\phi_{p,j}) = (b_i,\phi_p) \qquad p \in P, \tag{5.4}$$

$$\begin{cases} \dot{\sigma} = D\dot{\varepsilon}\ , & \dot{\varepsilon}^p = 0 \qquad \text{if } f(\sigma) < H(\bar{\varepsilon}^p)\ , \\ \dot{\sigma} = (D - D')\dot{\varepsilon}\ , & \dot{\varepsilon}^p = \frac{1}{H'}\partial f\partial f^*\dot{\sigma} \qquad \text{if } f(\sigma) = H(\bar{\varepsilon}^p) \text{ and } \partial f^*\dot{\sigma} \geq 0, \end{cases} \tag{5.5}$$

where D' denotes the matrix $D\partial f\partial f^* D/(H' + \partial f^* D\partial f)$.

The stress-strain relations (5.5) must be understood as follows: We apply the elastic rule $\dot{\sigma} = D\dot{\varepsilon}$ until $f(\sigma) = H(0)$ is satisfied at $t = t_0$, for instance. If $f(\sigma)$ is decreasing after t_0, then the element is still elastic after t_0. If it is increasing, the plastic stress - strain relation $\dot{\sigma} = (D - D')\dot{\varepsilon}$ must be applied, where $H' = H'(H^{-1}(f(\sigma)))$. The elastic state continues until $f(\sigma) = H(0)$ is satisfied again, and the plastic state continues until $\partial f^*\dot{\sigma} = 0$ is satisfied at $t = t_1$, for instance. In the latter case, assume that $f(\sigma)$ is decreasing after t_1. Then the next yield surface is $f(\sigma) = f(\sigma(t_1)) = H(\bar{\varepsilon}^p(t_1))$. This rule is applied to the subsequent yieldings and unloadings for all elements. Regarding the continuation of the solutions, it is now easy to prove

THEOREM 5.9. Assume that the solution of (5.4)~(5.5) exists in $(0,t_0)$ and that at $t = t_0$ some element satisfies $f(\sigma) = H(\bar{\varepsilon}^p)$. Then there is a positive constant δ such that in the interval $[t_0, t_0+\delta)$ the state of all the elements is definite; that is, it is either elastic or plastic. This determination depends on the data before t_0 and of b.

We omit the proof since it is almost the same as in the kinematic case, when we take into account that $H'(\bar{\varepsilon}^p) = H'(H^{-1}(f(\sigma)))$ must hold in the plastic state.

The energy estimates of the solution are easily obtained. Since in each element it holds that

(5.6) $$\dot{\varepsilon} - C\dot{\sigma} = 0 \text{ (elastic) or } \dot{\varepsilon} - C\dot{\sigma} = \frac{1}{H'}\,\partial f\,\partial f^*\dot{\sigma} \quad \text{(plastic)},$$

we have

$$\dot{\varepsilon}^*\sigma - (C\dot{\sigma})^*\sigma \;\left(- \frac{1}{H'}\,\partial f^*\sigma\,\partial f^*\dot{\sigma} \text{ for plastic element } \right) = 0.$$

Let Ω_p be the set of all plastic elements at time t. Integrating the above identity over e and summing on e, we have

$$(\dot{\varepsilon},\sigma) - (C\dot{\sigma},\sigma) - \int_{\Omega_p} \frac{1}{H'}\, f\partial f^*\dot{\sigma}\, dx = 0.$$

Since $1/H'\partial f^*\dot{\sigma} = \dot{\bar{\varepsilon}}^p$ in the plastic state and $\dot{\bar{\varepsilon}}^p = 0$ in the elastic state, we have

THEOREM 5.10. The following identity holds for all $t \in I$:

$$\frac{1}{2}\left(\|\dot{u}\|_\rho^2 + \|\sigma\|_C^2 \right) + \int_0^t\!\!\int_\Omega f\dot{\bar{\varepsilon}}^p\, dxdt = \int_0^t (b,\dot{u})dt.$$

For the higher derivatives we have

THEOREM 5.11. For any $t \in I$ it holds that

$$(5.7)\qquad \frac{1}{2}\left(\|\ddot{u}\|_\rho^2 + \|\dot{\sigma}\|_C^2 + \|\sqrt{H'}\dot{\varepsilon}^p\|^2\right) \le \int_0^t (\dot{b},\ddot{u})dt + \frac{1}{2}\left(\|\ddot{u}\|_\rho^2 + \|\dot{\sigma}\|_C^2\right)(0).$$

PROOF. Suppose that in a certain neighborhood of t there is no element which changes the state. Differentiating both sides of (5.6) we have either $\ddot{\sigma} = D\ddot{\varepsilon}$ or

$$\ddot{\varepsilon} - C\ddot{\sigma} - \left(\frac{d}{dt}\frac{1}{H'}\right)\partial f \partial f^* \dot{\sigma} - \frac{1}{H'}\left(\partial \dot{f} \partial f^* \dot{\sigma} + \partial f \frac{d}{dt}(\partial f^* \dot{\sigma})\right) = 0.$$

Multiplying both sides by $\dot{\sigma}$ and integrating with respect to t, we have by (5.4)

$$(5.8)\qquad \begin{aligned} &\frac{1}{2}\left[\|\ddot{u}\|_\rho^2 + \|\dot{\sigma}\|_C^2 + \int_{\Omega_p} \frac{1}{H'}(\partial f^* \dot{\sigma})^2 dx\right]_{t-\delta}^{t+\delta} \\ &+ \int_{t-\delta}^{t+\delta}\int_{\Omega_p}\left[\frac{1}{2}\left(\frac{d}{dt}\frac{1}{H'}\right)(\partial f^* \dot{\sigma})^2 + \frac{1}{H'}\partial \dot{f}^* \dot{\sigma} \partial f^* \dot{\sigma}\right] dxdt = \int_{t-\delta}^{t+\delta} (\dot{b},\ddot{u})dt, \end{aligned}$$

where δ is a small positive constant. Here we use the following relations:

(a) $\frac{d}{dt}\frac{1}{H'} = \frac{-H''}{(H')^2}\dot{\varepsilon}^p \ge 0,$

(b) $f \partial \dot{f}^* \dot{\sigma} = (S\dot{\sigma},\dot{\sigma}) - (\partial f^* \dot{\sigma})^2 \ge 0,$

(c) $\frac{1}{H'}(\partial f^* \dot{\sigma})^2 = H'(\dot{\varepsilon}^p)^2 = (\sqrt{H'}\,\dot{\varepsilon}^p)^2.$

Hence we have the following inequality.

$$(5.9)\qquad \frac{1}{2}\left[\|\ddot{u}\|_\rho^2 + \|\dot{\sigma}\|_C^2 + \|\sqrt{H'}\dot{\varepsilon}^p\|^2\right]_{t-\delta}^{t+\delta} \le \int_{t-\delta}^{t+\delta} (\dot{b},\ddot{u})dt.$$

Let us consider the e-part of the quantity in the blanket of (5.9) :

$$E^e(t) = \left(\|\ddot{u}\|_\rho^2 + \|\dot{\sigma}\|_C^2 + \|\sqrt{H'}\dot{\varepsilon}^p\|^2\right)_e(t).$$

We show below that this is non-increasing at the discontinuous point t_0, for instance, of $\dot{\sigma}$ and $\dot{\varepsilon}^p$. Consider the time intervals $I_i = (t',t_0)$ and $I_j = (t_0,t'')$ in which no change of state occurs for any element. There are three cases to be checked.

(a) The element e is elastic on I_i and plastic on I_j. In this case we have

$$E^e(t_0+0) = \|\ddot{u}\|_\rho^2(t_0+0) + \|\dot{\sigma}\|_C^2(t_0+0) + \|\frac{1}{\sqrt{H'}}\,\partial f^*\dot{\sigma}\|^2(t_0+0),$$

where the suffix e is omitted in the right side for simplicity's sake. Now e is plastic in I_j. Hence we have at $t = t_0+0$

$$(C\dot{\sigma},\dot{\sigma}) = (\dot{\varepsilon} - \frac{1}{H'}\,\partial f\partial f^*\dot{\sigma},\dot{\sigma})$$

$$= (\dot{\varepsilon},\dot{\sigma}) - (\frac{1}{H'}\partial f^*\dot{\sigma},\partial f^*\dot{\sigma}),$$

so that

$$E^e(t_0+0) = \|\ddot{u}\|_\rho^2(t_0+0) + (\dot{\varepsilon},(D-D')\dot{\varepsilon})(t_0+0)$$

$$\leq \|\ddot{u}\|_\rho^2(t_0-0) + (\dot{\varepsilon},D\dot{\varepsilon})(t_0-0)$$

$$= E^e(t_0-0),$$

since $\ddot{u}$ is continuous and D' is non-negative definite :

$$(D'\dot{\varepsilon},\dot{\varepsilon}) = (\frac{D\partial f\partial f^*D}{H'+\partial f^*D\partial f}\,\dot{\varepsilon},\dot{\varepsilon}) = (\frac{1}{H'+\partial f^*D\partial f}\,\partial f^*D\dot{\varepsilon},\partial f^*D\dot{\varepsilon}) \geq 0.$$

(b) The element is plastic in I_i and elastic in I_j. In this case $\partial f^*\dot{\sigma} = 0$ holds at $t = t_0+0$ as in the kinematic hardening case. Hence $E^e(t_0+0) = E^e(t_0-0)$.

(c) The case where t_0 is an accumulation point of the points at which the state of e changes. The situation is the same as in the kinematic hardening problem. Hence $E^e(t_0+0) = E^e(t_0-0)$ holds in this case too.

Thus $E^e(t)$ is non-increasing at any discontinuous point, and (5.7) follows by continuating the estimate (5.9). This completes the proof.

In conclusion, we have

THEOREM 5.12. Let $(u, \sigma, \bar{\varepsilon}^p)$ be the solution of (5.4)~(5.5). The following quantities are uniformly bounded on the triangulaion of Ω:

$$\|\dot{u}\| , \|\sigma\| , \|\ddot{u}\| , \|\dot{\sigma}\| , \|\dot{\bar{\varepsilon}}^p\| .$$

The weak representation of the present problem is as follows. Let us introduce a new variable $\hat{\varepsilon}^p$ and a new function G by

$$\hat{\varepsilon}^p = \int_0^{\bar{\varepsilon}^p} \sqrt{H'(\lambda)}\, d\lambda$$

$$G(\hat{\varepsilon}^p) = H(\bar{\varepsilon}^p).$$

We also define

$$B_G = \{ (\tau,\zeta) \in R^3 \times (-\delta^*,\infty) ; \quad f(\tau) \le G(\zeta) \}$$

$$K_G = \{ (\tau,\zeta) = (\sum_e \tau^e(t)\chi^e, \sum_e \zeta^e\chi^e) \in C^1_+(I) ; (\tau^e(t),\zeta^e(t)) \in B_G \} .$$

Let S^h be the set of functions $(u,\sigma,\hat{\varepsilon}^p) \in C^1_+(I)$ with the form

$$(u,\sigma,\hat{\varepsilon}^p) = (\sum_{p\in P} u^p(t)\phi_p(x), \sum_e \sigma^e(t)\chi^e, \sum_e \hat{\varepsilon}^e(t)\chi^e).$$

THEOREM 5.13. Problem (5.4)~(5.5) is equivalent to the following problem : Seek $(u,\sigma,\hat{\varepsilon}^p) \in S^h$ satisfying for all $t \in I$

(5.10) $$\rho(\ddot{u}_i,\phi_p) + \sum_j (\sigma_{ij},\phi_{p,j}) = (b_i,\phi_p) \qquad p \in P,$$

$$(5.11) \qquad (\dot{\varepsilon} - C\mathring{\sigma}, \tau - \sigma) - (\dot{\hat{\varepsilon}}^p, \zeta - \hat{\varepsilon}^p) \le 0 \qquad \text{for all } (\tau,\zeta) \in K_G,$$

and $(\sigma,\hat{\varepsilon}^p) \in K_G$, $u(0) = \sigma(0) = \hat{\varepsilon}^p(0) = 0$, $\dot{u}(0) = a$.

PROOF. Consider an arbitrary element e at time t. If e is elastic

$$f(\sigma) < H(\bar{\varepsilon}^p) = G(\hat{\varepsilon}^p); \text{ that is, } \quad (\sigma,\hat{\varepsilon}^p) \in B_G.$$

Also it holds that

$$(5.12) \qquad \dot{\varepsilon} - C\dot{\sigma} = 0, \quad \dot{\hat{\varepsilon}}^p = \sqrt{H'(\bar{\varepsilon}^p)}\,\dot{\bar{\varepsilon}}^p = 0.$$

On the other hand, if e is plastic, then

$$f(\sigma) = H(\bar{\varepsilon}^p) = G(\hat{\varepsilon}^p);$$

that is, $(\sigma,\hat{\varepsilon}^p)$ lies on the boundary of B_G. Also, since

$$\begin{pmatrix} \dot{\varepsilon} - C\dot{\sigma} \\ \\ -\dot{\hat{\varepsilon}}^p \end{pmatrix} = \begin{pmatrix} \dot{\varepsilon}^p \\ \\ -\sqrt{H'(\bar{\varepsilon}^p)}\,\dot{\bar{\varepsilon}}^p \end{pmatrix} = \begin{pmatrix} \partial f \\ \\ -G'(\hat{\varepsilon}^p) \end{pmatrix} \dot{\bar{\varepsilon}}^p,$$

the vector in the left side is parallel to the outward normal to B_G at the boundary point $(\sigma,\hat{\varepsilon}^p)$ (see Fig. 9). Hence (5.11) follows. Since the solution of (5.10)~(5.11) is unique, as is easily proved, the proof is complete.

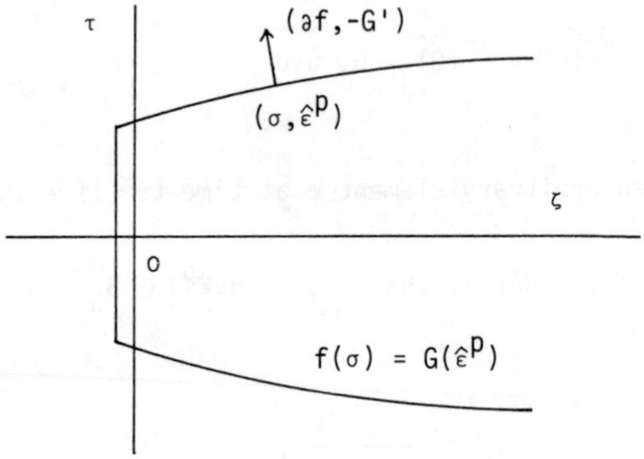

Fig. 9 The normality condition

CHAPTER 6

TWO-DIMENSIONAL QUASI-STATIC SEMIDISCRETE SYSTEM

6.1. Semidiscrete finite element approximation

In this chapter we extend the results of Chapter 4 to the general quasi-static problem. Since we can treat both the kinematic and isotropic cases in the same manner, we discuss mainly the former problem.

The displacements are approximated by u_i $(i = 1,2)$ of the form

$$u_i(t) = \sum_{p \in P} u_i^p(t)\phi_p(x),$$

and $\{u_i^p(t)\}$ are determined by the following system of equations:

$$(1.1) \qquad \sum_{j=1}^{2} (\sigma_{ij}, \phi_{p,j}) = (b_i, \phi_p) \qquad p \in P,$$

$$(1.2) \qquad \dot{\sigma} = D\dot{\varepsilon}\ ,\ \dot{\alpha} = 0 \qquad \text{if } f(\sigma - \alpha) < z_0 \text{ or } f(\sigma - \alpha) = z_0 \text{ and } \partial f * \dot{\sigma} < 0,$$

$$(1.3) \qquad \dot{\sigma} = (D - D')\dot{\varepsilon}\ ,\ \dot{\alpha} = (\sigma - \alpha)\frac{\partial f * \dot{\sigma}}{f} \text{ if } f(\sigma - \alpha) = z_0 \text{ and } \partial f * \dot{\sigma} \geq 0.$$

This system is the same as (1.5)~(1.7) of Chapter 5 except the " or " part and the inertia term. We seek $C_+^1(I)$- solution (u,σ,α) satisfying zero initial conditions.

6.2. Determination of the first derivatives

We say that an element e is elastic (or plastic) if (1.2) (or (1.3)) is satisfied on this element. Let E denote the set of all elements of Ω .

Since we started from zero initial condition, all elements of E are elastic until some element satisfies $f(\sigma) = z_0$ at $t = t_0$, for example. It is clear that our problem has a unique piecewise analytic solution $(u, \sigma, 0)$ in the interval $(0,t_0)$. Let E_0 be the set of all the elements which satisfy $f(\sigma) = z_0$ at $t = t_0$. $E - E_0$ is clearly still elastic after t_0, since the solution must be continuous. Hence the next question is whether the elements of E_0 yield at $t = t_0$ or still remain elastic.

The key is to guess the sign of $\partial f^*\dot{\sigma}$ at $t= t_0+ 0$, since if it is positive (or negative) then the stress point σ moves outside (or inside) the yield surface $f(\sigma) = z_0$. We note that the signs of $\partial f^*\dot{\sigma}$ and $\partial f^*D\dot{\varepsilon}$ for $t > t_0$ are the same, since

$$\partial f^*\dot{\sigma} = \partial f^*D\dot{\varepsilon} \quad \left(= \partial f^*D\dot{\varepsilon}\left(1 - \frac{\partial f^*D\partial f}{\eta+\partial f^*D\partial f}\right) \right) \quad \text{if elastic (if plastic).}$$

Thus let us consider the following system, which must be satisfied by the first derivatives of the solution (if it exists) :

$$(2.1) \qquad \sum_{j=1}^{2} (\sigma^0_{ij},\phi_{p,j}) = (\dot{b}_i(t_0+0),\phi_p) \qquad p \in P,$$

$$(2.2) \qquad \sigma^0 = D\varepsilon^0 \qquad \text{for } E - E_0,$$

$$(2.3) \quad \begin{cases} \sigma^0 = D\varepsilon^0 & \text{in } D_- = \{ u^0 \ ; \ \partial f^*(t_0)D\varepsilon^0 < 0 \} \\ \sigma^0 = (D - D')\varepsilon^0 & \text{in } D_+ = \{ u^0 \ ; \ \partial f^*(t_0)D\varepsilon^0 \geq 0 \} \end{cases}$$

for $E - E_0$, where

$$\varepsilon^0 = \varepsilon(u^0) \qquad u^0_i = \sum_{p\in P} u^{p,0}_i\phi_p \ (i = 1,2)$$

$$D' = D'(t_0).$$

THEOREM 6.1. Problem (2.1)~(2.3) has a unique solution (u^0,σ^0). This

u^o minimizes the functional

$$F_1(u^o) = \frac{1}{2}(\sigma^o, \varepsilon^o) - (\dot{b}(t_0+0), u^o)$$

under the subsidiary conditions (2.2) and (2.3).

PROOF. (i). Let $(\ , \ )_e$ be the $L^2(e)$ inner product of (vector) functions ($e \in E$), and define

$$F^e = \frac{1}{2}(\sigma^o, \varepsilon^o)_e - (\dot{b}(t_0+0), u^o)_e.$$

Then F^e is a C^1-class function of u^o_i and thus $F_1 = \sum F^e$ is too. This is clear for $e \in E - E_0$. Let e be an element of E_0. We have in D_-

$$\frac{\partial F^e}{\partial u_i^{p,o}} = (D\varepsilon^o, \frac{\partial \varepsilon^o}{\partial u_i^{p,o}})_e - (\dot{b}(t_0+0), \frac{\partial u^o}{\partial u_i^{p,o}})_e$$

and similarly in D_+

$$\frac{\partial F^e}{\partial u_i^{p,o}} = ((D - D')\varepsilon^o, \frac{\partial \varepsilon^o}{\partial u_i^{p,o}})_e - (\dot{b}(t_0+0), \frac{\partial u^o}{\partial u_i^{p,o}})_e.$$

These derivatives coincide on the plane $X_e = \{ u^o \ ; \ \partial f^*(t_0) D\varepsilon^o = 0 \}$, since

$$D'\varepsilon^o = \frac{D\partial f \partial f^* D}{\eta + \partial f^* D \partial f}\varepsilon^o = 0$$

on this plane. This proves the C^1-continuity of F^e.

(ii). $D - D'$ is positive definite in the sense that there is a positive constant C depending only on η such that

(2.4) $$((D - D')\varepsilon^o, \varepsilon^o)_{R^3} \geq C\|\varepsilon^o\|^2_{R^3}.$$

To see this, we first note that the vector ∂f is bounded. In fact, by an elementary calculation we have

$$\frac{\sqrt{2}}{2} \leq \|\partial f\|_{R^3} \leq \sqrt{3}.$$

Now, since D is symmetric and positive definite, there are constants C_1 and C_2 such that

$$(2.5) \qquad C_1 \leq \partial f^* D \partial f \leq C_2.$$

Now let λ and x be a positive eigenvalue and the corresponding eigenvector of $\sqrt{D}\partial f \partial f^* \sqrt{D}^*$ ($D = \sqrt{D}^* \sqrt{D}$) ; that is,

$$\sqrt{D}\partial f \partial f^* \sqrt{D}^* x = \lambda x.$$

Multiplying both sides by $\partial f^* \sqrt{D}^*$, we have

$$\partial f^* D \partial f \partial f^* \sqrt{D}^* x = \lambda \partial f^* \sqrt{D}^* x.$$

Since $\partial f^* \sqrt{D}^* x \neq 0$, we have $\lambda = \partial f^* D \partial f$. Hence we have

$$\begin{aligned}(D\partial f \partial f^* D\varepsilon^0, \varepsilon^0) &= (\sqrt{D}\partial f \partial f^* \sqrt{D}^* \sqrt{D}\, \varepsilon^0, \sqrt{D}\varepsilon^0) \\ &\leq \| \sqrt{D}\partial f \partial f^* \sqrt{D}^* \| \, (\sqrt{D}\varepsilon^0, \sqrt{D}\varepsilon^0) \\ &\leq \partial f^* D \partial f \, (D\varepsilon^0, \varepsilon^0),\end{aligned}$$

and by (2.5)

$$\begin{aligned}((D - D')\varepsilon^0, \varepsilon^0) &= ((D - \frac{D\partial f \partial f^* D}{\eta + \partial f^* D \partial f})\varepsilon^0, \varepsilon^0) \\ &\geq (1 - \frac{\partial f^* D \partial f}{\eta + \partial f^* D \partial f})(D\varepsilon^0, \varepsilon^0) \\ &\geq \frac{\eta}{\eta + C_2} (D\varepsilon^0, \varepsilon^0) \geq C \|\varepsilon^0\|^2_{R^3},\end{aligned}$$

which proves (2.4). Therefore $F_1(u^0)$ is bounded below by Korn's inequality (we do not need this inequality as far as we treat the semidiscrete system). And so $F_1(u^0)$ has a minimum point which is also stationary . However, if u^0 is a stationary point of F_1, then at this point $\sigma^0 = \sigma(u^0)$ determined by (2.2) or (2.3) must satisfy the stationary condition which is equivalent to

(2.1). Hence the problem (2.1)~(2.3) has a solution which minimizes F_1.

(iii). To prove the uniqueness of the solution, it suffices to show the uniqueness of the stationary point of $F_1(u^0)$. For each element $e \in E_0$ we consider the hyperplane X_e in u^0-space. Let $\{R_\lambda\}$ be the partition of the u^0-space by these planes. In each R_λ the $\sigma^0 - \varepsilon^0$ relation is definite for all elements of E. Also, in R_λ, $F_1(u^0)$ is a positive definite quadratic form of u^0 and the stationary point is at most one.

Now assume that there are two stationary points, $u^1 \in R_\lambda$ and $u^2 \in R_\mu (\lambda \neq \mu)$. Consider the line

$$L \quad : u^1 + t(u^2 - u^1) \qquad t \in [0,1].$$

This line goes through at least two regions of $\{R_\lambda\}$ when t moves from 0 to 1. Then the function

$$g(t) = F_1(u^1 + t(u^2 - u^1))$$

is smooth and non-degenerate quadratic on t in each region. Therefore, if u^1 is a stationary point (that is, if $F_1(u^1)$ is the minimum) then g(t) must be strictly increasing in [0,1], which contradicts that u^2 is another stationary point. This completes the proof.

We want to show that the solution (u^0,σ^0) of the problem (2.1)~(2.3) is the first derivative at $t = t_0 + 0$ of the solution of (1.1)~(1.3), provided the latter has a solution. By Theorem 6.1 we can determine the sign of $\partial f^* D\varepsilon^0$ at $t = t_0$ for the elements of E_0. We denote by E^e and E^p the sets of all elements of E_0 for which this sign is negative and nonnegative, respectively, and solve the following initial value problem set up at $t = t_0$:

$$(2.6) \qquad \sum_{j=1}^{2} (\sigma_{ij}, \phi_{p,j}) = (b_i, \phi_p) \qquad p \in P$$

$$(2.7)\quad \begin{cases} \dot\sigma = D\dot\varepsilon, \quad \dot\alpha = 0 & \text{for } E - E^p \\ \dot\sigma = (D - D')\dot\varepsilon, \quad \dot\alpha = (\sigma - \alpha)\dfrac{\partial f^*\dot\sigma}{f} & \text{for } E^p, \end{cases}$$

where $\dot\varepsilon = \varepsilon(\dot u)$, $D' = D'(t)$, $f = f(\sigma - \alpha)$ and $(u,\sigma,\alpha)(t_0) = (u,\sigma,\alpha)(t_0 - 0)$.

THEOREM 6.2. The initial value problem (2.6)~(2.7) has a unique analytic solution (u,σ,α) in a certain neighborhood of t_0, and $(\dot u,\dot\sigma)(t_0 + 0) = (u^0,\sigma^0)$.

PROOF. Differentiate both sides of (2.6) with respect to t and denote the resulting equation by $(2.6)^{(1)}$. Substituting the $\dot\sigma$ - $\dot\varepsilon$ relation of (2.7) into $(2.6)^{(1)}$ and solving the resulting equation with respect to $\dot u$, we have $\dot u$ as an analytic function of σ, α and t in a certain neighborhood of t_0. Therefore (2.7) can be regarded as a system of ordinary differential equations of the form

$$\frac{d}{dt}\begin{pmatrix}\sigma\\ \alpha\end{pmatrix} = X(\sigma,\alpha,t)$$

where X is an analytic vector function. This system has a unique analytic solution under the given initial condition. Hence the problem (2.6)~(2.7) has a unique analytic solution (u, σ, α) in a certain neighborhood of t_0. Furthermore, the solution (u^0, σ^0) of the problem (2.1)~(2.3) satisfies the following problem (A) at $t = t_0 + 0$ when we put $(\dot u,\dot\sigma)(t_0+0) = (u^0,\sigma^0)$:

$$(A)\quad \sum_{j=1}^{2} (\dot\sigma_{ij},\phi_{p,j}) = (\dot b_i,\phi_p) \qquad p \in P$$

$$\begin{cases} \dot\sigma = D\dot\varepsilon & \text{for } E - E^p \\ \dot\sigma = (D - D')\dot\varepsilon & \text{for } E^p. \end{cases}$$

Let E_1 be the set of all elements of E^p for which $\partial f^*(t_0)D\dot\varepsilon(t_0 + 0) = 0$ holds for the solution of (2.6)~(2.7). If E_1 is empty, then the next state is completely determined for all the elements. Notice that the elements of $E - E_0$ are still elastic after t_0, and for those of E^e it holds that

$$f^2(\sigma) - f^2(\sigma(t_0)) = 2\int_{t_0}^{t_0+\delta} f(\sigma)\partial f^*\dot{\sigma}\, ds < 0 \qquad (\delta > 0)$$

for small δ, and for those of E^p we have

$$\partial f^*_{(\sigma-\alpha)}\dot{\sigma} = \partial f^*_{(\sigma-\alpha)}D\dot{\varepsilon}\ (1-\theta) > 0 \quad \text{and } f(\sigma-\alpha) = z_0$$

for a while after t_0, where $0 < \theta < 1$. In other words, the $\sigma - \varepsilon$ relation of the elements of E^e and of E^p are already chosen correctly. We emphasize here that this determination is dependent only on the data at $t = t_0 - 0$ and the given function b.

If E_1 is not empty, however, we have to guess the sign of $d/dt(\partial f^*\dot{\sigma})$ at $t = t_0 + 0$. In this case the following theorem is important. Replace some elements of E_1 from E^p to E^e and solve the initial value problem (2.6)~(2.7) for this new E^p. Let the new system be denoted by (2.6)'~(2.7)'. Then this problem has a unique solution (u, σ, α) under the same initial condition at $t = t_0 + 0$. Moreover, we have

THEOREM 6.3. For any element of E, the value $(\dot{u},\dot{\sigma},\dot{\alpha})(t_0 + 0)$ is determined independently of the choice of the next $\sigma - \varepsilon$ relation of E_1.

PROOF. Consider the following problem at $t = t_0 + 0$:

(A)' $$\sum_j (\dot{\sigma}_{ij}, \phi_{p,j}) = (\dot{b}_i, \phi_p) \qquad p \in P$$

$$\begin{cases} \dot{\sigma} = D\dot{\varepsilon} & \text{for } E - (\text{new})E^p \\ \dot{\sigma} = (D - D')\dot{\varepsilon} & \text{for } (\text{new})E^p. \end{cases}$$

Since the solution (u^0,σ^0) of (2.1)~(2.3) satisfies $D'(t_0)\varepsilon^0 = 0$ for the elements of E_1, (u^0,σ^0) is also a solution of(A)' at $t=t_0+0$. However, the solution of (A)' is unique with respect to $(\dot{u},\dot{\sigma})(t_0+0)$, and so $(\dot{u},\dot{\sigma})(t_0+0) = (u^0,\sigma^0)$. From this the theorem follows, since $\dot{\alpha}(t_0+0) = 0$ for E_1.

It is thus assured that the elements of $E - E_1$ behave so as to satisfy the subsidiary conditions of (1.1) for any choice of the next state of E_1. In other words, the next state of the elements of $E - E_1$ is already determined. Hence we can exclude them from our consideration.

6.3. Determination of the higher derivatives

It is clear that the next state of the elements of E_1 is determined by the sign of $d/dt(\partial f^*\dot{\sigma})$ at $t = t_0+0$. It is easy to see that this sign and that of $d/dt(\partial f^*D\dot{\varepsilon})$ at $t = t_0+0$ are the same for the elements of E_1. Hence we consider the following equations which must be satisfied by the second derivatives of the solution.

(3.1) $$\sum_j (\sigma_{ij},\phi_{p,j}) = (\frac{d^2}{dt^2} b_i(t_0+0),\phi_p) \qquad p \in P$$

(3.2) $$\begin{cases} \sigma^0 = D\varepsilon^0 & \text{for the elastic element of } E - E_1 \\ \sigma^0 = [(D - D')\varepsilon^0 - (\frac{d}{dt} D')\dot{\varepsilon}]_{t_0+0} & \text{for the plastic element of } E - E_1 , \end{cases}$$

and for the element of E_1

(3.3) $$\begin{cases} \sigma^0 = D\varepsilon^0 & \text{in } D_- = \{u^0; \partial f^*(t_0)D\varepsilon^0 + r_1 < 0\} \\ \sigma^0 = [(D - D')\varepsilon^0 - (\frac{d}{dt} D')\dot{\varepsilon}]_{t_0+0} & \text{in } D_+ = \{u^0; \partial f^*(t_0)D\varepsilon^0 + r_1 \geq 0\}, \end{cases}$$

where ε^0, u^0, ∂f and D' are defined similarly as before, and

$$r_1 = [(\frac{d}{dt} \partial f)^* D\dot{\varepsilon}]_{t_0+0}$$

Note that $(\dot{u},\dot{\sigma},\dot{\alpha})(t_0+0)$ is the derivative of the solution of (2.6)~(2.7) and is already known independently of the next state of E_1.

THEOREM 6.4. Problem (3.1)~(3.3) has a unique solution (u^0,σ^0). This u^0 minimizes the following functional F_2 under the subsidiary conditions (3.2) and (3.3).

$$F_2(u^0) = \frac{1}{2}(\sigma^0,\varepsilon^0) - \frac{1}{2}(\lambda_2,\bar{\varepsilon}) - (\frac{d^2}{dt^2} b(t_0+0),u^0), \tag{3.4}$$

where $\lambda_2 = [dD'/dt\cdot\dot{\varepsilon}]_{t_0+0}$ and

$$\bar{\varepsilon} = \begin{cases} 0 \ (\text{ or } \varepsilon^0) & \text{for the elastic (or plastic) element of } E - E_1 \\ \left.\begin{array}{ll} \varepsilon^0_* & \text{in } D_- \\ \varepsilon^0 & \text{in } D_+ \end{array}\right\} & \text{for the element of } E_1. \end{cases}$$

Here ε^0_* is an arbitrary fixed vector included in the hyperplane of u^0 - space

$$X_e : \{ u^0 \ ; \ \partial f^*(t_0)D\varepsilon^0 + r_1 = 0 \}. \tag{3.5}$$

PROOF. (i). $F_2(u^0)$ is a continuous function. To prove this, let F^e be the e - part of F_2 as before ($e \in E$). For $e \in E - E_1$ the continuity of F^e is clear. For $e \in E_1$ the discontinuity of F^e might appear across the plane X_e. However, at $t = t_0+0$ it holds that on X_e

$$D'\varepsilon^0 + (\frac{d}{dt} D')\dot{\varepsilon} = \frac{D\partial f\partial f^* D}{\eta+\partial f^* D\partial f}\varepsilon^0 + \frac{D\partial f(d\partial f/dt)^* D}{\eta+\partial f^* D\partial f}\dot{\varepsilon} \tag{3.6}$$

$$= \frac{D\partial f}{\eta+\partial f^* D\partial f}(\partial f^* D\varepsilon^0 + r_1) = 0.$$

Hence σ^0 is continuous with respect to u^0, and so is the first term of F^e

The jump at X_e of the second term of F^e is $1/2(\lambda_2, \varepsilon_*^0 - \varepsilon^0)_e$. But if ε_*^0 and ε^0 belong to X_e, then $D'(\varepsilon_*^0 - \varepsilon^0) = 0$ at $t = t_0 + 0$ by (3.6). Therefore, since D' is symmetric, we have at $t = t_0 + 0$

$$(\lambda_2, \varepsilon_*^0 - \varepsilon^0)_e = ((\frac{d}{dt} D')\dot{\varepsilon}, \varepsilon_*^0 - \varepsilon^0)_e = - (D'\varepsilon^0, \varepsilon_*^0 - \varepsilon^0)_e$$

$$= - (\varepsilon^0, D'(\varepsilon_*^0 - \varepsilon^0))_e = 0.$$

This implies the continuity of the second term of F^e and hence of F^e itself.

(ii). $F_2(u^0)$ is a C^1-class function of u^0. To prove this, we check three cases. First, let e be an elastic element of $E - E_1$. Then F^e is smooth, since

$$F^e = \frac{1}{2}(D\varepsilon^0, \varepsilon^0)_e - (\frac{d^2}{dt^2} b(t_0+0), u^0)_e.$$

Secondly, let e be the plastic element of $E - E_1$. Then

$$F^e = \frac{1}{2}(\sigma^0, \varepsilon^0)_e - \frac{1}{2}([(\frac{d}{dt} D')\dot{\varepsilon}](t_0+0), \varepsilon^0)_e - (\frac{d^2}{dt^2} b(t_0+0), u^0)_e.$$

Therefore

$$\begin{aligned} \frac{\partial F^e}{\partial u_i^{p,0}} &= ((D - D')\varepsilon^0, \frac{\partial \varepsilon^0}{\partial u_i^{p,0}})_e - ([(\frac{d}{dt} D')\dot{\varepsilon}](t_0+0), \frac{\partial \varepsilon^0}{\partial u_i^{p,0}})_e \\ &\quad - (\frac{d^2}{dt^2} b(t_0+0), \frac{\partial u^0}{\partial u_i^{p,0}})_e \qquad (3.7) \\ &= (\sigma^0, \frac{\partial \varepsilon^0}{\partial u_i^{p,0}})_e - (\frac{d^2}{dt^2} b(t_0+0), \frac{\partial u^0}{\partial u_i^{p,0}})_e. \end{aligned}$$

Since σ^0 is continuous, this equality shows the smoothness of F^e. Finally, let e be an element of E_1. Then the following is clear in D_- :

$$\frac{\partial F^e}{\partial u_i^{p,0}} = (\sigma^0, \frac{\partial \varepsilon^0}{\partial u_i^{p,0}})_e - (\frac{d^2}{dt^2} b(t +0), \frac{\partial u^0}{\partial u_i^{p,0}})_e.$$

On the other hand, the relation (3.7) is valid in D_+. Hence, the continuity of $F_2(u^0)$ has been proved.

(iii). $F_2(u^0)$ is a positive definite (piecewise) quadratic form which is bounded below. Hence the minimizing point u^0 of F_2 exists, and (u^0,σ^0) where σ^0 is determined by (3.2)~(3.3) is a solution of (3.1). These are proved in the same way as in Theorem 6.1. This completes the proof of the theorem.

The solution (u^0,σ^0) of the problem (3.1)~(3.3) is the second derivative of the true solution in the following sense. Divide $E_1 = E_1^e + E_1^p$, where

$$E_1^e = \{e \in E_1 \ ; \ \partial f^*(t_0)D\varepsilon^0 + r_1 < 0\ \}$$

$$E_1^p = \{e \in E_1 \ ; \ \partial f^*(t_0)D\varepsilon^0 + r_1 \geq 0\ \},$$

and solve (2.6)~(2.7) replacing E^p with the new $E^p = E^p - E_1^e$. Let (u, σ, α) be its solution. By theorem 6.3, $(\dot{u},\dot{\sigma},\dot{\alpha})(t_0+0)$ is the same as the first derivatives of the solution of the problem (2.6)~(2.7) with old E^p. Moreover we have

THEOREM 6.5. Let (u^0,σ^0) be the solution of (3.1)~(3.3).

(1) $(u^0,\sigma^0) = (\ddot{u},\ddot{\sigma})(t_0+0)$.

(2) Let E_2 be the set of elements of E_1^p such that

$$\partial f^*(t_0)D\varepsilon^0 + r_1 = 0. \tag{3.8}$$

Then for every element of E, $(\ddot{u},\ddot{\sigma},\ddot{\alpha})(t_0+0)$ is determined independently of the choice of the next σ - ε relation of E_2.

The proof is omitted since it is the same as that of Theorem 6.3. Note that for the element of E_2 it holds that

$$[D'\varepsilon^0 + (\frac{d}{dt}D')\dot{\varepsilon}]_{t_0+0} = 0$$

where (u^0,σ^0) is the solution of (3.1) ~ (3.3).

Furthermore, if E_2 is not empty, we repeat this argument until E_k becomes empty for a certain $k = K < \infty$. It should happen that there are some elements for which the equality

$$[\frac{d^k}{dt^k}(\partial f^* \dot{\sigma})]_{t_0+0} = 0$$

holds for all k, then we assign the plastic $\sigma - \varepsilon$ relation to these elements with the same reason as before.

For the completeness, we decribe below the procedure to determine the derivatives of order $k + 1$ of the solution at $t = t_0+0$ when E_k ($k \geq 2$) is not empty. Assume that the derivatives of order less than or equal to k of the solution are already determined independently of the choice of the next $\sigma - \varepsilon$ relation of E_k :

$$E_k = \{ e \in E_{k-1} ; [\frac{d^{k-1}}{dt^{k-1}}(\partial f^* D\dot{\varepsilon})]_{t_0+0} = 0 \}.$$

Let us define (formally)

$$\lambda_{k+1} = [\frac{d^k}{dt^k}(D'\dot{\varepsilon}) - D'\frac{d^k}{dt^k}\dot{\varepsilon}]_{t_0+0}$$

$$r_k = [\frac{d^k}{dt^k}(\partial f^* D\dot{\varepsilon}) - \partial f^* D\frac{d^k}{dt^k}\dot{\varepsilon}]_{t_0+0} .$$

Let (u^0, σ^0) be the solution of the following problem set up at $t = t_0 + 0$:

(3.9) $$\sum_j (\sigma^0_{ij}, \phi_{p,j}) = (\frac{d^{k+1}}{dt^{k+1}} b_i, \phi_p) \qquad p \in P$$

(3.10) $$\begin{cases} \sigma^0 = D\varepsilon^0 & \text{for the elastic element of } E - E_k \\ \sigma^0 = (D - D')\varepsilon^0 - \lambda_{k+1} & \text{for the plastic element of } E - E_k , \end{cases}$$

and for the elements of E_k

(3.11) $$\begin{cases} \sigma^0 = D\varepsilon^0 & \text{in } D_- = \{ u^0 ; \partial f^*(t_0)D\varepsilon^0 + r_k < 0 \} \\ \sigma^0 = (D - D')\varepsilon^0 - \lambda_{k+1} & \text{in } D_+ = \{ u^0 ; \partial f^*(t_0)D\varepsilon^0 + r_k \geq 0 \}. \end{cases}$$

THEOREM 6.6. The problem (3.9)~(3.11) has a unique solution (u^0,σ^0). This u^0 minimizes the following functional F_{k+1} under the conditions (3.10) and (3.11).

$$F_{k+1}(u^0) = \frac{1}{2}(\sigma^0,\varepsilon^0) - \frac{1}{2}(\lambda_{k+1},\bar{\varepsilon}) - (\frac{d^{k+1}}{dt^{k+1}} b(t_0+0),u^0)$$

where

$$\bar{\varepsilon} = \begin{cases} 0 \ (\text{ or } \varepsilon^0) & \text{for the elastic (or plastic) element of } E - E_k \\ \left.\begin{array}{ll} \varepsilon_*^0 & \text{in } D_- \\ \varepsilon^0 & \text{in } D_+ \end{array}\right\} & \text{for the element of } E_k , \end{cases}$$

and ε_*^0 is an arbitrary vector in u^0 - space included in the hyperplane

$$X_e \ : \{ u^0 \ ; \ \partial f^*(t_0)D\varepsilon^0 + r_k = 0 \}.$$

Now classify as $E_k = E_k^e + E_k^p$, where $\partial f^*D\varepsilon^0 + r_k$ is negative for E_k^e and nonnegative for E_k^p at $t = t_0+ 0$, and solve (2.6)~(2.7) replacing E^p of the preceding stage with the new $E^p = E^p - E_k^e$. Let (u,σ,α) be its solution. Then (u^0,σ^0) is the derivative of order k + 1 of (u,σ) at $t_0+ 0$. Also, the derivative of order k + 1 of (u,σ,α) is determined independently of the choice of the next σ - ε relation of $E_{k+1} = \{ e \in E_k \ ;[\frac{d^k}{dt^k}(\partial f^*D\dot{\varepsilon})]_{t_0+0} = 0 \}$.

Summarizing the above results, we have

THEOREM 6.7. The σ - ε relation of each element of E is determined uniquely after t_0, and the problem (1.1)~(1.3) has a unique solution in a certain time interval $I_\delta = [t_0, t_0+ \delta)$ $(\delta > 0)$, which is analytic in I_δ .

So far we have discussed only the case of the initial yielding. The above method and results are valid almost word for word for the subsequent yieldings and unloadings, as it was in the dynamic problems. Since the boundedness of the solution is assured by the energy estimate, we can continuate the solution over the whole interval I. In fact, we have

THEOREM 6.8. Let $(u, \sigma, \alpha) \in C^1_+$ be the solution of (1.1)~(1.3) in an interval $I' \subset I$. Then there is a constant C which is independent of I' such that

$$\|\dot{u}\|, \|\dot{\varepsilon}\|, \|\dot{\sigma}\|, \|\dot{\alpha}\| \leq C. \tag{3.12}$$

PROOF. In I' we have $(\dot{\sigma}, \dot{\varepsilon}) = (\dot{b}, \dot{u})$. Therefore it holds that

$$((D - D')\dot{\varepsilon}, \dot{\varepsilon}) \leq |(\dot{b}, \dot{u})| \leq \|\dot{b}\| \|\dot{u}\| \leq C \|\dot{b}\| \tag{3.13}$$

The boundedness of the first three quantities of (3.12) thus follows from the positivity of the matrix D - D'. To estimate the last quantity, use the following relation which holds for any t in I' :

$$\dot{\varepsilon} - D^{-1}\dot{\sigma} = \frac{1}{\eta} S\dot{\alpha},$$

where $f(\sigma)\,\partial f(\sigma) = S\sigma$.

The conclusion of this chapter is

THEOREM 6.9. There is a unique solution $(u, \varepsilon, \sigma, \alpha) \in C^1_+(I)$ of the problem (1.1)~(1.3).

PROOF. The continuation of the solution over the whole interval I is now evident even when the number of the yieldings and unloadings is infinite. Hence the existence of the solution is assured. And though the above argument also shows the uniquenes, we shall prove it by another method.

Let $K = K_\alpha$ ($\alpha \in C^1_+(I)$) be defined by

$$K = \{\tau \in C^1_+(I) \ ; \ \ f(\tau - \alpha) \leq z_0 \ \}.$$

If $(u,\varepsilon,\sigma,\alpha)$ satisfies (1.1)~(1.3), then $\sigma \in K$ and the following hold :

$$\text{(3.14)} \qquad \sum_{j=1}^{2} (\sigma_{ij},\phi_{p,j}) = (b_i,\phi_p) \qquad p \in P$$

$$\text{(3.15)} \qquad \int_I (\dot{\varepsilon} - C\dot{\sigma},\tau - \sigma)dt \leq 0 \qquad \text{for all } \tau \in K$$

$$\text{(3.16)} \qquad \dot{\alpha} = \eta\, S^{-1}(\dot{\varepsilon} - C\dot{\sigma}\ ) \qquad \text{a.e. } I \qquad (C = D^{-1}).$$

Assume that $(u,\varepsilon,\sigma,\alpha)_*$ too satisfies (1.1). Since $\sigma \in K$ can be written as $\sigma = \alpha + \theta z_0$, where $f(\theta) \leq 1$, we have

$$\int_I (\dot{\varepsilon} - C\dot{\sigma},\alpha + \sigma_* - \alpha_* - \sigma)dt \leq 0$$

$$\int_I (\dot{\varepsilon}_* - C\dot{\sigma}_*,\ \alpha_* + \sigma - \alpha \ - \ \sigma_*)dt \leq 0.$$

Define $(U,E,\Sigma,A)=(u,\varepsilon,\sigma,\alpha) - (u,\varepsilon,\sigma,\alpha)_*$. Adding these inequalities, we have

$$0 \geq \int_I (\dot{E} - C\dot{\Sigma},A - \Sigma)dt$$

$$= \frac{1}{\eta}\int_I (S\dot{A},A)dt + \int_I (C\dot{\Sigma},\Sigma)dt,$$

from which the uniqueness follows.

6.4. Isotropic hardening problem

The variational mechanism which determined the yielding and unloading in the kinematic hardening problem works in the isotropic hardening problem too. In this case, the unknowns $\{u^p_i(t)\}$ ($p \in P$, $i=1,2$) are determined by the following system of equations :

$$(4.1) \qquad \sum_{j=1}^{2} (\sigma_{ij}, \phi_{p,j}) = (b_i, \phi_p) \qquad p \in P$$

$$(4.2) \qquad \begin{cases} \dot{\sigma} = D\dot{\varepsilon} & \text{if } f(\sigma) < H(\bar{\varepsilon}^p), \text{ or } f(\sigma) = H(\bar{\varepsilon}^p) \text{ and } \partial f^*\dot{\sigma} < 0 \\ \dot{\sigma} = (D - D')\dot{\varepsilon} & \text{if } f(\sigma) = H(\bar{\varepsilon}^p) \text{ and } \partial f^*\dot{\sigma} \geq 0, \end{cases}$$

where

$$\bar{\varepsilon}^p = \int_0^t \frac{\sigma^*\dot{\varepsilon}^p}{f(\sigma)}\,dt \qquad \dot{\varepsilon}^p = \dot{\varepsilon} - C\dot{\sigma}$$

$$D' = \frac{D\partial f \partial f^* D}{H' + \partial f^* D \partial f} \qquad H' = H'(H^{-1}(f(\sigma))).$$

Assume that we start from $t = 0$ and that at $t = t_0$ some element satisfies $f(\sigma) = H(0)$ at the first time. Let E_0 be the set of elements for which this equality is satisfied. To guess the sign of $\partial f^*\dot{\sigma}$ at $t = t_0 + 0$, we consider the system of equations which correspond to (2.1)~(2.3). Then Theorem 6.1 holds without any modification. The solution (u^0, σ^0) of this system is the first derivative at $t = t_0 + 0$ of the solution (u, σ) of the following initial value problem set up at $t = t_0$ (see also Theorem 6.2) :

$$(4.3) \qquad \sum_{j=1}^{2} (\sigma_{ij}, \phi_{p,j}) = (b_i, \phi_p) \qquad p \in P$$

$$(4.4) \qquad \begin{cases} \dot{\sigma} = D\dot{\varepsilon}, \quad \dot{\varepsilon}^p = 0 & \text{for } E - E^p \\ \dot{\sigma} = (D - D')\dot{\varepsilon}, \quad \dot{\varepsilon}^p = \frac{1}{H'}\partial f \partial f^*\dot{\sigma} & \text{for } E^p, \end{cases}$$

where E^p is defined in the same way as before. (Note that for the element of E^p, $f(\sigma) = H(\bar{\varepsilon}^p)$ holds automatically, so long as $\partial f^*\dot{\sigma} \geq 0$ is satisfied. See §1 of Chapter 1.) Now let E_1 be the set of the elements of E^p for which $\partial f^*\dot{\sigma}|_{t_0+0} = 0$ holds for the solution of (4.3)~(4.4). Then the next state of the element of $E - E_1$ can be determined correctly, and it is independent of the choice of the next state of the elements of E_1 (that is, Theorem 6.3 is valid). This argument can be repeated until the next state of all the elements of E is determined correctly . In fact, the functionals F_k (k = 1.2,..) to determine the derivatives of the solution are exactly the same as before. Hence we can continuate the solution across $t = t_0$ and as far as the solution remains bounded.

As the a priori estimates of the solution, we have

THEOREM 6.10. There is a constant C which is independent of h such that

(4.5) $$\|\dot{\sigma}\|,\ \|\dot{\varepsilon}\|,\ \|\dot{\bar{\varepsilon}}^p\| \leq C\ \|\dot{b}\| .$$

PROOF. The boundedness of $\dot{\sigma}$ and $\dot{\varepsilon}$ is obvious (see the proof in the kinematic hardening case). Now in the plastic state, it holds that

$$\partial f * \dot{\sigma} = H'\dot{\bar{\varepsilon}}^p.$$

Since H' is bounded below by the assumption, and $\|\partial f\|$ is bounded, we have the desired estimates.

Now introduce a new variable $\hat{\varepsilon}^p$ and a function G as in the dynamic problem :

$$\hat{\varepsilon}^p = \int_0^{\bar{\varepsilon}^p} \sqrt{H'(\lambda)}\ d\lambda \quad ,$$

$$G(\hat{\varepsilon}^p) = H(\bar{\varepsilon}^p).$$

THEOREM 6.11. The problem (4.1)~(4.2) is equivalent to the following problem : Seek $(u,\sigma,\hat{\varepsilon}^p) \in S^h$ which satisfies for all $t \in I$

(4.6) $$\sum_j (\sigma_{ij},\phi_{p,j}) = (b_i,\phi_p) \qquad p \in P,$$

(4.7) $(\dot{\varepsilon} - C\dot{\sigma}, \tau - \sigma) - (\dot{\hat{\varepsilon}}^p, \zeta - \hat{\varepsilon}^p) \leq 0$ for all $(\tau,\zeta) \in K_G$,

and $(\sigma,\hat{\varepsilon}^p) \in K_G$, $u(0) = \sigma(0) = \hat{\varepsilon}^p(0) = 0$.

For the definitions of S^h and K_G , and the proof of the theorem, see Theorem 5.13 and its proof.

CHAPTER 7

NUMERICAL STABILITY IN DYNAMIC ELASTIC-PLASTIC PROBLEMS

7.1. Finite difference approximation of the acceleration

As a preliminary study of the fully discrete system to approximate the elastic-plastic dynamic problems, we consider the following finite difference approximation to the equation (1.1) of Chapter 1 :

$$(1.1) \qquad \rho\, D_t D_{\bar{t}} u_n + \sigma_n = 0,$$

where D_t and $D_{\bar{t}}$ are the forward and backward difference operators with time mesh Δt. In so far as we treat " geometrically linear " problems, the outer force b does not play any essential role, so we neglect this term for the time being.

To determine the yielding and unloading in the present problem, we put the following criterion : Let $P^{\pm}$ and E be the lines and the elastic zone defined as §1 of Chapter 2 :

$$P^{\pm} \; ; \; \sigma \mp z_0 = (1 - \xi)k(u \mp \frac{1}{k} z_0)$$

$$E = \{(u,\sigma) \; ; \; (1 - \xi)ku - \xi z_0 < \sigma < (1 - \xi)ku + \xi z_0 \}.$$

(A) The yielding. If $(u_{n-1},\sigma_{n-1}) \in \bar{E}$ and $(u_n,\tilde{\sigma}_n) \notin \bar{E}$ for $\tilde{\sigma}_n = \sigma_{n-1} + k(u_n - u_{n-1})$, then determine $(\bar{u}_n,\bar{\sigma}_n) \in P^+ \cup P^-$ by

$$\bar{\sigma}_n = \sigma_{n-1} + \theta_n(\tilde{\sigma}_n - \sigma_{n-1}) \qquad 0 \le \theta_n < 1$$

$$\bar{u}_n = u_{n-1} + \frac{1}{k}(\bar{\sigma}_n - \sigma_{n-1}),$$

and use the modified plastic rule

$$\sigma_n - \bar{\sigma}_n = (1 - \xi)k(u_n - \bar{u}_n).$$

For the subsequent steps, the usual plastic rule $D_t\sigma_n = (1-\xi)kD_tu_n$ is applied as long as $D_tu_n \geq 0$ (or ≤ 0) is satisfied for $(u_n, \sigma_n) \in P^+$ (or $\in P^-$).

(B) The unloading. If $D_tu_n < 0$ (or > 0) is satisfied for $(u_n, \sigma_n) \in P^+$(or $\in P^-$),then the elastic rule is applied on and after σ_{n+1}; that is, $D_t\sigma_n = kD_tu_n$.

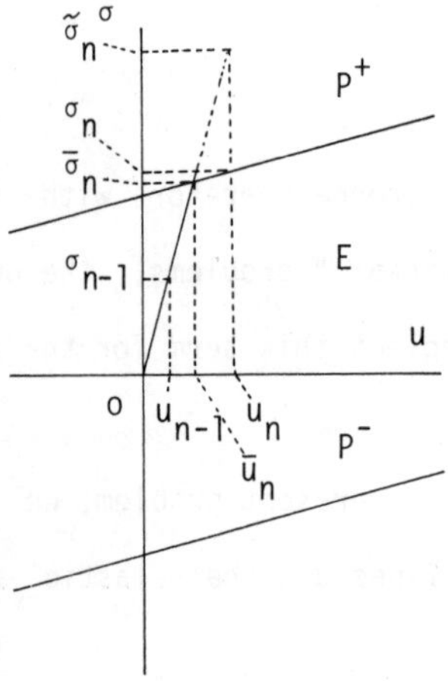

Fig. 10 The yielding

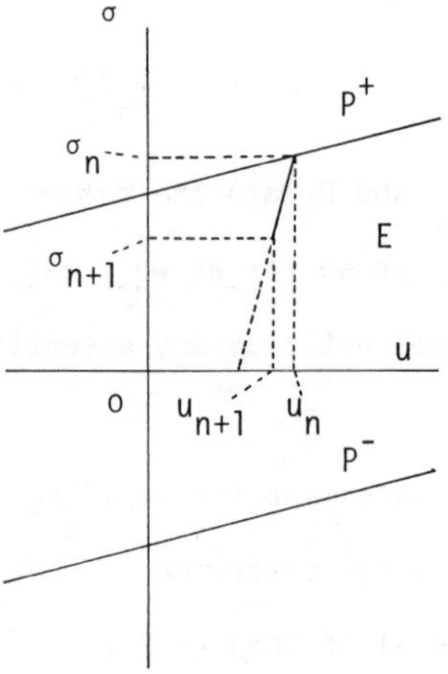

Fig. 11 The unloading

By step(n) we denote the step to determine σ_n. Let $u_{(j)}$ be the displacement when (j+1)-th change of the state occurs (hence $u_{(j)} = u_n$ or $= \bar{u}_n$ for some n). We say that stage(m) begins from step(n) if $u_{n-1} \leq u_{(m)} < u_n$ in the case of yielding, and if $u_{(m)} = u_{n-1}$ in the case of unloading.

THEOREM 7.1. Under the above yielding-unloading rule, the equation (1.1) is represented at step(n) in stage(m) as follows.

$$\rho D_t D_{\bar{t}} u_n + k u_n - \xi k \sum_{j=0}^{m} (-1) (u_n - u_{(j)}) = 0. \tag{1.2}$$

PROOF. Assume m = 0. If stage(0) begins from step(p), then, since $\bar{\sigma}_p = k u_{(0)}$, we have

$$\sigma_n = \bar{\sigma}_p + (1 - \xi) k (u_n - u_{(0)}) \qquad n \geq p$$

$$= k u_n - \xi k (u_n - u_{(0)}).$$

Hence the theorem is correct for m = 0. Assume next that it is correct until m = r.

(i). The case where stage(r) is elastic : Assume that stage(r+1) begins from step(p). We then have

$$\sigma_n = \bar{\sigma}_p + (1 - \xi) k (u_n - u_{(r+1)}) \qquad n \geq p. \tag{1.3}$$

Since $\bar{\sigma}_p - \sigma_{p-1} = k(u_{(r+1)} - u_{p-1})$ and

$$\sigma_{p-1} = k u_{p-1} - \xi k \sum_{j=0}^{r} (-1)^j (u_n - u_{(j)}),$$

(1.3) implies, taking into account that r is odd, that

$$\sigma_n = k u_n - \xi k \sum_{j=0}^{r+1} (-1) (u_n - u_{(j)}).$$

Hence (1.2) holds for m = r+1.

(ii). The case where stage(r) is plastic : Assume that stage(r+1) begins from step(p). Since

$$\sigma_n = \sigma_{p-1} + k (u_n - u_{p-1}) \qquad n \geq p,$$

$$\sigma_{p-1} = k u_{p-1} - \xi k \sum_{j=0}^{r} (-1)^j (u_{p-1} - u_{(j)}),$$

$u_{p-1} = u_{(r+1)}$ and r is even, we have

$$\sigma_n = ku_{p-1} - \xi k u_{p-1} - \xi k \sum_{j=0}^{r} (-1)^j(-u_{(j)}) + k(u_n - u_{r+1})$$
$$= ku_n - \xi k \sum_{j=0}^{r+1} (-1)^j(u_n - u_{(j)}).$$

Hence (1.2) holds for m = r+1. This completes the proof.

An energy inequality similar to that in Theorem 2.5 holds for the present system. To see this, we introduce the quantity

$$E^{(m)}(n) = \rho(D_t u_n)^2 + ku_n u_{n+1} - \xi k \sum_{j=0}^{m} (-1)^j(u_n - u_{(j)})(u_{n+1} - u_{(j)})$$
$$+ \frac{\xi k}{(1-\xi)} u_{(0)}^2.$$

Then multiplying both sides of (1.2) by $(D_t + D_{\bar{t}})u_n$, we have

(1.4) $$D_{\bar{t}}E^{(m)}(n) = 0$$

as far as stage(m) continues. In order to connect $E^{(m)}$ with $E^{(m-1)}$ we check the following two cases :

(A). The case of unloading. If the velocity changes the sign at u_n, then $u_{(m)} = u_n$, so that

$$E^{(m)}(n) = \rho(D_t u_n)^2 + ku_n u_{n+1} - \xi k \sum_{j=0}^{m-1} (-1)^j(u_n - u_{(j)})(u_{n+1} - u_{(j)})$$
$$+ \frac{\xi k}{1-\xi} u_{(0)}^2$$
$$= E^{(m-1)}(n).$$

Hence $E^{(m)}$ and $E^{(m-1)}$ coincide when the unloading occurs.

(B). The case of yielding. Assume that $u_{(m)}$ lies between u_n and u_{n+1}. Since m is even, we then have

$$E^{(m)}(n) = \rho(D_t u_n)^2 + k u_n u_{n+1} - \xi k \sum_{j=0}^{m} (-1)^j (u_n - u_{(j)})(u_{n+1} - u_{(j)}) + \frac{\xi k}{1-\xi} u_{(0)}^2$$

$$= E^{(m-1)}(n) - \xi k (u_n - u_{(m)})(u_{n+1} - u_{(m)})$$

$$\leq E^{(m-1)}(n) + \xi k \Delta t^2 (D_t u_n)^2 .$$

Notice that the above $E^{(m-1)}(n)$ can also be written as

$$E^{(m-1)}(n) = (\rho - \frac{k}{2}\Delta t^2)(D_t u_n)^2 + \frac{k}{2}(u_n^2 + u_{n+1}^2) - \xi k \sum_{j=0}^{m-1} (-1)^j (u_n - u_{(j)})(u_{n+1} - u_{(j)}) + \frac{\xi k}{1-\xi} u_{(0)}^2 .$$

It is easy to see that this quantity is positive under the condition

(1.5) $$\rho - \frac{k}{2}\Delta t^2 > 0,$$

and larger than $(\rho - k/2\ \Delta t^2)(D_t u_n)^2$. Hence, under this condition, we have the following relation connecting these two stages ;

(1.6) $$E^{(m)}(n) \leq [1 + \xi k\ \Delta t^2/(\rho - \frac{k}{2}\Delta t^2)\ E^{(m-1)}(n).$$

In conclusion of the above discussion we have

THEOREM 7.2. Under the condition (1.5), the following estimate holds.

$$E^{(m)}(n) \leq e^{\xi k \Delta t \cdot T/(\rho - k\Delta t^2/2)} \left[(\rho - \frac{k}{2}\Delta t^2)(D_t u_0)^2 + \frac{k}{2}(u_0^2 + u_1^2) \right]$$

$$(n\Delta t \leq T).$$

REMARK. Condition (1.5) is regarded as a stability condition. In fact, a discrete scheme introduced later to approximate plane stress problems is stable (in the energy sense) under a similar condition. Note that the stability condition for the elastic vibrations, generally, takes the form (1.5). Hence we can say that the plasticity has no influence on the numerical stability as far as the stability in energy is concerned.

7.2. Numerical stability (1)

We want to apply the above idea of finding a stability condition to the following system of difference equations which approximate the equations (1.1) of Chapter 3 :

$$\rho_i D_t D_{\bar{t}} u_i(n) + \sigma_i(n) - \sigma_{i+1}(n) = 0 \qquad (i = 1 \sim N), \tag{2.1}$$

where

$$D_{\bar{t}}\sigma_i(n) = \begin{cases} k_i D_{\bar{t}} U_i(n) & \text{for the elastic spring} \\ (1 - \xi_i) k_i D_{\bar{t}} U_i(n) & \text{for the plastic spring,} \end{cases}$$

and $U_i(n) = u_i(n) - u_{i-1}(n)$, $u_0(n) = k_{N+1} = 0$. The yielding and unloading are defined in the same way as before (see Fig. 12 below). In the present case, $\sigma_i(n)$ and $U_i(n)$ stand for σ_n and u_n.

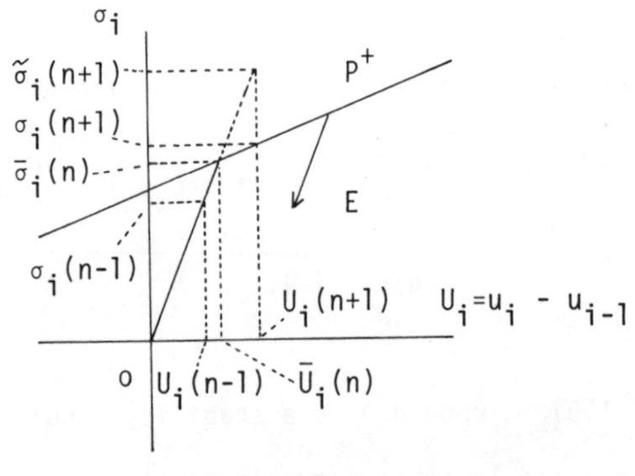

Fig. 12 Yielding and unloading for the multiple system

It is easy to prove the next theorem.

THEOREM 7.3. When i-th spring remains stage(m_i), the above difference scheme is represented as follows :

$$(2.2)\qquad \rho_i D_t D_{\bar t} u_i(n) + [k_i U_i(n) - \xi_i k_i \sum_{j=0}^{m_i} (-1)^j (U_i(n) - U_i^{(j)})] - [k_{i+1} U_{i+1}(n) - \xi_{i+1} k_{i+1} \sum_{j=0}^{m_{i+1}} (-1)^j (U_{i+1}(n) - U_{i+1}^{(j)})] = 0 \qquad (i = 1 \sim N),$$

where $U_i^{(j)}$ denotes the value of U_i at which the i-th spring changes the state and stands for $u_{(j)}$ of the single system.

Let $m = (m_1, m_2, \ldots, m_N)$ and introduce the quanity

$$E^{(m)}(n) = \sum_i [\rho_i (D_t u_i(n))^2 - \tfrac{1}{2} k_i \Delta t^2 (D_t U_i(n))^2 + \tfrac{1}{2}(U_i^2(n) + U_i^2(n+1)) - \xi_i k_i \sum_{j=0}^{m_i} (-1)^j (U_i(n) - U_i^{(j)})(U_i(n+1) - U_i^{(j)}) + \frac{\xi_i k_i}{1-\xi_i}(U_i^{(0)})^2]$$

By step(n) we denote the step to determine $\sigma(n)$.

THEOREM 7.4. Assume that the i-th spring is in stage(m_i) at step(n). If

$$(2.3)\qquad \beta = \operatorname{Min}_i [\rho_i - (k_i + k_{i+1}) \Delta t^2] > 0,$$

then the following estimate holds for $n \le T/\Delta t$:

$$(2.4)\qquad E^{(m)}(n) \le e^{T\gamma\Delta t} E(0) \qquad \text{where } \gamma = \operatorname{Max} \frac{\xi_i k_i + \xi_{i+1} k_{i+1}}{\rho_i - (k_i + k_{i+1})\Delta t^2}$$

PROOF. It is easy to prove the following three identities :

(A). $\rho_i D_t D_{\bar t} u_i(n)\,(D_t + D_{\bar t}) u_i(n) = D_{\bar t}(\rho_i (D_t u_i(n))^2)$,

$$\text{(B).}\quad \sum_i (k_iU_i(n) - k_{i+1}U_{i+1}(n))(D_t + D_{\bar{t}})u_i(n)$$

$$= \sum_i (k_iU_i(n) - k_{i+1}U_{i+1}(n))(D_t + D_{\bar{t}})U_i(n)$$

$$+ \sum_i (k_iU_i(n) - k_{i+1}U_{i+1}(n))(D_t + D_{\bar{t}})u_{i-1}(n)$$

$$= \sum_i k_iU_i(n)(D_t + D_{\bar{t}})U_i(n)$$

$$= D_{\bar{t}} \sum_i k_iU_i(n)\, U_i(n+1)$$

$$= \frac{1}{2} D_{\bar{t}} \sum_i k_i[U_i^2(n) + U_i^2(n+1) - (D_tU_i(n))^2\Delta t^2],$$

$$\text{(C).}\quad \xi_ik_i \sum_{j=0}^{m_i} (-1)^j(U_i(n) - U_i^{(j)})(D_t + D_{\bar{t}})u_i(n)$$

$$= \xi_ik_i \sum_{j=0}^{m_i} (-1)^j(U_i(n) - U_i^{(j)})(D_t + D_{\bar{t}})U_i(n)$$

$$+ \xi_ik_i \sum_{j=0}^{m_i} (-1)^j(U_i(n) - U_i^{(j)})(D_t + D_{\bar{t}})u_{i-1}(n).$$

Now multiply (2.2) by $(D_t + D_{\bar{t}})u_i(n)$ and add on i. Then by (A),(B),(C) and the identity

$$\sum_i [\xi_ik_i \sum_{j=0}^{m_i} (-1)^j(U_i(n) - U_i^{(j)}) - \xi_{i+1}k_{i+1} \sum_{j=0}^{m_{i+1}} (-1)^j(U_{i+1}(n) - U_{i+1}^{(j)})$$

$$\times (D_t + D_{\bar{t}})u_i(n)\ ]$$

$$= \sum_i [\xi_ik_i \sum_{j=0}^{m_i} (-1)^j(U_i(n) - U_i^{(j)})](D_t + D_{\bar{t}})U_i(n)$$

$$= D_{\bar{t}} \sum_i [\xi_ik_i \sum_{j=0}^{m_i} (-1)^j(U_i(n) - U_i^{(j)})(U_i(n+1) - U_i^{(j)})],$$

$E^{(m)}(n)$ is constant as far as stage(m_i) ($i = 1 \sim N$) continues.

In order to examine the continuity of E(n), we consider the i-th spring. Assume that this spring is in stage(m_i) and is plastic. If stage(m_i) begins from step(n') and there is no change of the state of other spring until step(n), we then set $m' = (m_1, m_2, .., m_{i-1} .., m_i - 1, .., m_N)$. Clearly it holds that

$$E^{(m)}(n) = E^{(m)}(n'-1) = E^{(m')}(n'-1) - \xi_i k_i (-1)^{m_i} (U_i(n') - U_i^{(m_i)})(U_i(n'-1) - U_i^{(m_i)})$$

$$\leq E^{(m')}(n'-1) + \frac{\xi_i k_i}{2} \Delta t^2 (D_t U_i(n'-1))^2.$$

Hence, even if all the springs change their state at one time, it holds that

$$(2.5) \qquad E^{(m)}(n) \leq E^{(m')}(n'-1) + \sum_i \frac{\xi_i k_i}{2} \Delta t^2 (D_t U_i(n'-1))^2$$

for a suitable choice of m'. Since

$$\sum_i [\rho_i (D_t u_i)^2 - \frac{k_i}{2} \Delta t^2 (D_t U_i)^2] \geq \sum_i [(\rho_i - (k_i + k_{i+1})\Delta t^2)(D_t u_i)^2]$$

and

$$\sum_i \frac{\xi_i k_i}{2} \Delta t^2 (D_t U_i(n'-1))^2 \leq [\operatorname*{Max}_i \frac{(\xi_i k_i + \xi_{i+1} k_{i+1})\Delta t^2}{\rho_i - (k_i + k_{i+1})\Delta t^2}] E(n'-1),$$

(2.4) follows from (2.5). Note that $E^{(m)}(n)$ is positive under the condition (2.3). This completes the proof.

7.3. Numerical stability (2)

In this section we consider an explicit finite difference approximation of the system (1.5)~(1.7) of Chapter 5. The main purpose of the analysis is to prove that the approximate scheme is stable under the same condition which gurantees the stability of the system when plasticity is not considered.

We seek the approximate value of u_i at time step n in the form

$$u_i(n) = \sum_{p \in P} u_i^p(n) \phi_p.$$

The explicit scheme considered in this section is based on the central difference approximation of the inertia terms :

$$(3.1) \qquad (\rho D_t D_{\bar{t}} u_i(n), \phi_p) + \sum_j (\sigma_{ij}(n), \phi_{p,j}) = 0 \qquad p \in P.$$

As the discrete initial conditions, we assume $u_i^p(0) = 0$, $u_i^p(1) = \Delta t\ a_i(p)$. The strain-displacement relation is the same as before.

To introduce a discrete stress-strain relation, we choose an element e arbitrarily and fix it.

DEFINITION.

(1). By the yield surface of step(0) of an element e, we mean the ellipsoid in R^3 defined by

$$\{\ \tau \in R^3\ ;\ f(\tau - \alpha(0)) = z_0\}.$$

(2). ELASTIC RULE. We say that $\{\sigma(n+1), \alpha(n+1)\}$ is determined by the elastic rule, if $\sigma(n)$ lies in or on the yield surface of step(n), and if $\{\ \sigma(n+1), \alpha(n+1)\ \}$ and the yield surface of step(n+1) are determined by the following rule :

$$(3.2)_a \qquad D_t\varepsilon(n) = CD_t\sigma(n),$$

$$(3.2)_b \qquad D_t\alpha(n) = 0,$$

$(3.2)_c$ the equation defining the yield surface of step(n+1) is

$$f(\tau - \alpha(n+1)) = \text{Max}\ (z_0, \underset{m\leq n}{\text{Max}}\ [f(\sigma(m+1) - \alpha(m+1))]).$$

PLASTIC RULE. We say that $\{\sigma(n+1), \alpha(n+1)\}$ is determined by the plastic rule if $\sigma(n)$ is on the yield surface of step(n), and if $\{\sigma(n+1), \alpha(n+1)\}$ and the yield surface of step(n+1) are determined by the following rule :

$$(3.3)_a \qquad D_t\varepsilon(n) = \partial f_n \frac{\partial f_n^* D_t\sigma(n)}{n} + CD_t\sigma(n),$$

$$(3.3)_b \qquad D_t\alpha(n) = (\ \sigma(n) - \alpha(n)\ )\ \frac{\partial f_n^* D_t\sigma(n)}{f_n} \qquad (f_n = f(\sigma_n - \alpha_n)\ ),$$

$(3.3)_c$ the equation defining the yield surface of step(n+1) is

$$f(\tau - \alpha(n+1)) = f(\sigma(n+1) - \alpha(n+1)).$$

ELASTIC-PLASTIC RULE. Assume that $\sigma(n)$ lies in the yield surface of step(n) and that the point $\sigma(n+1)$ determined by the elastic rule from $\{\sigma(n), \alpha(n)\}$ is outside the yield surface of step(n). If $\partial f_n^*(\tilde{\sigma}(n+1) - \sigma(n)) \geq 0$, then $\{\sigma(n+1),\alpha(n+1)\}$ is determined by the following rule :

Choose θ_n (> 0) so that the point

$$\bar{\sigma}(n) = \sigma(n) + \theta_n(\tilde{\sigma}(n+1) - \sigma(n))$$

comes on the yield surface of step(n). Define

$$\bar{f}_n = f|_{(\sigma,\alpha)=(\bar{\sigma}(n),\alpha(n))}$$

$$\partial\bar{f}_n = \partial f|_{(\sigma,\alpha)=(\bar{\sigma}(n),\alpha(n))}.$$

Set $\bar{\varepsilon}(n) = \varepsilon(n) + C(\bar{\sigma}(n) - \sigma(n))$. Determine $\{\sigma(n+1), \alpha(n+1)\}$ by

$(3.4)_a$ $$\varepsilon(n+1) - \bar{\varepsilon}(n) = \partial\bar{f}_n \frac{\partial\bar{f}_n^*(\sigma(n+1)-\bar{\sigma}(n))}{\eta} + C(\sigma(n+1) - \bar{\sigma}(n)),$$

$(3.4)_b$ $$D_t\alpha(n) = (\bar{\sigma}(n) - \alpha(n)) \frac{\partial\bar{f}_n^*(\sigma(n+1)-\bar{\sigma}(n))}{\Delta t\bar{f}_n}.$$

$(3.4)_c$ The equation defining the yield surface of step(n+1) is

$$f(\tau - \alpha(n+1)) = f(\sigma(n+1) - \alpha(n+1)).$$

REMARK. Define $\bar{D}'(n)$ by

$$\bar{D}'(n) = \frac{D\partial\bar{f}_n\partial\bar{f}_n^*D}{\eta+\partial\bar{f}_n^*D\partial\bar{f}_n}.$$

Then $(3.4)_a$ is written also as follows :

$$(3.4)'_a \qquad D_t\sigma(n) = [D - (1-\theta_n)\bar{D}'(n)]D_t\varepsilon(n).$$

To see this, invert $(3.4)_a$ to get

$$\sigma(n+1) - \bar{\sigma}(n) = (D - \bar{D}'(n))(\varepsilon(n+1) - \bar{\varepsilon}(n)).$$

Substituting $\bar{\sigma}(n) - \sigma(n) = D(\bar{\varepsilon}(n) - \varepsilon(n))$ into this, we have

$$(3.5) \qquad \sigma(n+1) - \sigma(n) = D(\varepsilon(n+1) - \varepsilon(n)) - \bar{D}'(n)(\varepsilon(n+1) - \bar{\varepsilon}(n)).$$

We also have

$$\varepsilon(n+1) - \varepsilon(n) = C(\tilde{\sigma}(n+1) - \sigma(n)) = \frac{1}{\theta_n} C(\bar{\sigma}(n) - \sigma(n))$$
$$= \frac{1}{\theta_n}(\bar{\varepsilon}(n) - \varepsilon(n)),$$

so that

$$\varepsilon(n+1) - \bar{\varepsilon}(n) = (1-\theta_n)(\varepsilon(n+1) - \varepsilon(n)).$$

Substituting this into (3.5) we have $(3.4)'_a$. The cases $\theta_n = 1$ and $\theta_n = 0$ stand for the elastic and plastic rules, respectively.

REMARK. When the elastic-plastic rule is applied, the vector $\tilde{\sigma}(n+1) - \bar{\sigma}(n)$ is transversal to the yield surface at the stress point $\bar{\sigma}(n)$. This is the case also for the vector $\sigma(n+1) - \bar{\sigma}(n)$, since

$$\operatorname{sgn} \partial\bar{f}_n^*(\tilde{\sigma}(n+1) - \bar{\sigma}(n)) = \operatorname{sgn} \partial\bar{f}_n^*(\tilde{\sigma}(n+1) - \sigma(n))$$
$$= \operatorname{sgn} \partial\bar{f}_n^* D D_t\varepsilon(n) = \operatorname{sgn} \partial\bar{f}_n^*[D - \frac{D\partial\bar{f}_n\partial\bar{f}_n^* D}{\eta + \partial\bar{f}_n^* D\partial\bar{f}_n}]D_t\varepsilon(n)$$
$$= \operatorname{sgn} \partial\bar{f}_n^*[D - \bar{D}'(n)]D_t\varepsilon(n)$$
$$= \operatorname{sgn} \partial\bar{f}_n^*[D - \bar{D}'(n)](\varepsilon(n+1) - \bar{\varepsilon}(n))$$

$= \mathrm{sgn}\ \partial \bar{f}^*_n(\sigma(n+1) - \bar{\sigma}(n)).$

The relation between the stress and strain increments is given as follows :

DISCRETE STRESS-STRAIN RELATION (see Fig. 13).

We start from $u(0) = \sigma(0) = \alpha(0) = 0$.

(A). Determine $\{\tilde{\sigma}(n+1), \tilde{\alpha}(n+1)\}$ ($n \geq 0$) by the elastic rule as long as $\sigma(n)$ is in the yield surface of step(n). If $\tilde{\sigma}(n+1)$ is still in the yield surface of of step(n), we then define

$$\{\sigma(n+1), \alpha(n+1)\} = \{\tilde{\sigma}(n+1), \tilde{\alpha}(n+1)\} .$$

(B). If $\sigma(n)$ is in the yield surface of step(n), and if $\tilde{\sigma}(n+1)$ determined by the elastic rule comes on or outside the yield surface of step(n),then $\{\sigma(n+1), \alpha(n+1)\}$ is determined by

$(B)_a$ the elastic-plastic rule if $\partial f^*_n(\tilde{\sigma}(n+1) - \sigma(n)) \geq 0$, and otherwise by

$(B)_b$ the elastic rule.

The subsequent relations are given by the following procedure :

(C). If $\sigma(n+1)$ is on the yield surface of step(n+1), then determine $\{\tilde{\sigma}(n+2), \tilde{\alpha}(n+2)\}$ by the plastic rule.

$(C)_a$ If $\partial f^*_{n+1}(\tilde{\sigma}(n+2)-\sigma(n+1)) > 0$, then $\{\sigma(n+2), \alpha(n+2)\} = \{\tilde{\sigma}(n+2), \tilde{\alpha}(n+2)\}$.

Otherwise

$(C)_b$ determine $\{\sigma(n+2), \alpha(n+2)\}$ anew by the elastic rule.

(D). If $\sigma(n+1)$ remains in the yield surface of step(n+1), then return to procedure (A) $\rightarrow$ (B), replacing n by n+1.

REMARK. The elastic rule is applied in the following three cases :

(1). $\sigma(n)$ is in the yield surface of step(n), and $\tilde{\sigma}(n+1)$ determined by the elastic rule is too.

(2). $\sigma(n)$ is on the yield surface of step(n) and $\sigma(n+1)$ determined by the

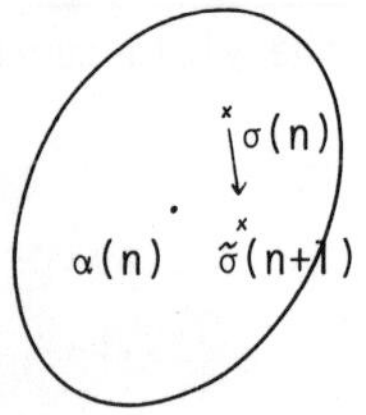

elastic rule (1)

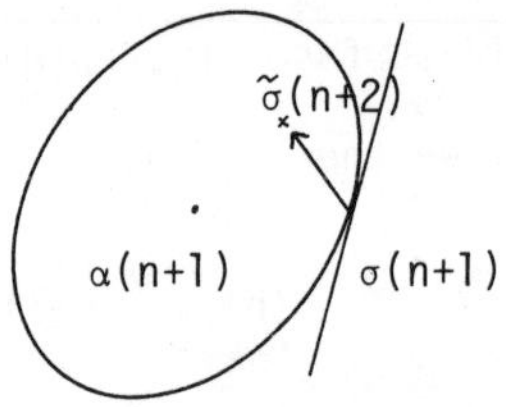

elastic rule (2)

(unloading)

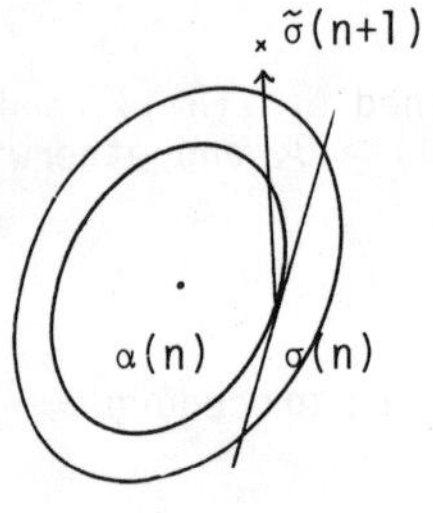

elastic rule (3)

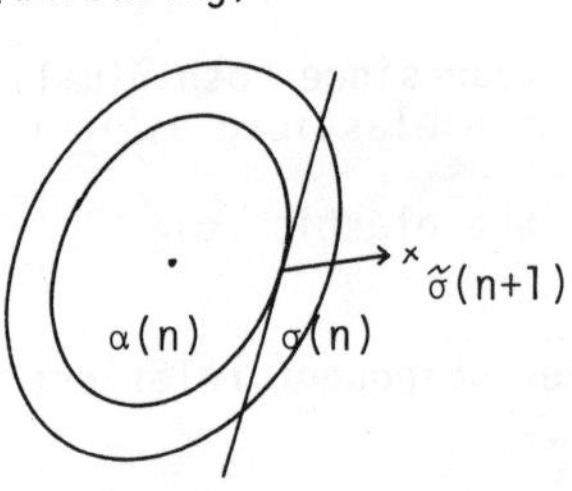

elastic-plastic rule

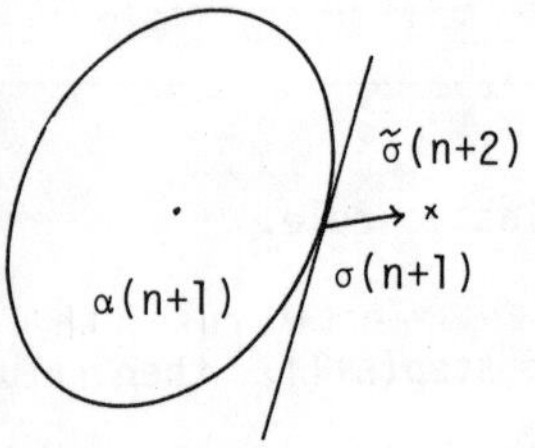

plastic rule

Fig. 13 Discrete stress-strain relations

plastic rule satisfies

$$\partial f_n^* (\tilde{\sigma}(n+1) - \sigma(n)) < 0. \tag{3.6}$$

This case corresponds to the unloading.

(3). $\sigma(n)$ is in the yield surface of step(n), and $\tilde{\sigma}(n+1)$ determined by the elastic rule goes out of this yield surface. Nevertheless, the following inequality holds :

$$\partial f_n^* (\tilde{\sigma}(n+1) - \sigma(n)) < 0.$$

REMARK. Suppose that $\{\tilde{\sigma}(n+1), \tilde{\alpha}(n+1)\}$ is determined by the plastic rule and that condition (3.6) holds. Then $\{\sigma(n+1),\alpha(n+1)\}$ must be determined anew by the elastic rule. In this case, (3.6) again holds. That is

$$\partial f_n^*(\sigma(n+1) - \sigma(n)) < 0.$$

This is clear since both $\tilde{\sigma}(n+1)$ and $\sigma(n+1)$ are determined by $\varepsilon(n+1)$, and

$$\partial f_n^*(\tilde{\sigma}(n+1) - \sigma(n)) = \partial f_n^* D D_t \varepsilon(n)\left(1 - \frac{\partial f_n^* D \partial f_n}{\eta + \partial f_n^* D \partial f_n}\right)\Delta t$$

in plastic state. This relation does not hold in general for the quasi-static problems.

Let $\{u(n),\sigma(n),\alpha(n)\}$ $(n \geq 0)$ be the solution obtained by the above method. We seek a criterion to ensure the stability of this solution. To do this, we first provide

THEOREM 7.5. (1). If $\sigma(n+1)$ is determined by the plastic rule, then

$$\partial f_n^*(D_t\sigma(n) - D_t\alpha(n)) = 0.$$

(2). If $\sigma(n+1)$ is determined by the elastic-plastic rule, then

$$\partial\bar{f}_n^*(\sigma(n+1) - \bar{\sigma}(n) - [\alpha(n+1) - \alpha(n)]) = 0,$$

where $\bar{\sigma}(n)$ and $\partial\bar{f}_n$ are those defined before.

(3). The yield surface expands monotonically.

(4). Let f' and $\partial f'$ be the values of f and ∂f at $\{\sigma, \alpha\} = \{\sigma', \alpha'\}$, respectively and set $k = f/f'$. Then

$$(\sigma - \alpha)^* \partial f' = k(\sigma' - \alpha')^* \partial f.$$

(5). For any $n \geq 0$ it holds that

$$D_t\varepsilon(n) = \frac{1}{\eta} SD_t\alpha(n) + CD_t\sigma(n).$$

PROOF. Since (1),(2) and (5) are easy to prove, we show (3) and (4). During the elastic deformation, the center of the yield surface is fixed, and the surface itself is also fixed or simply expands. When the plastic rule is applied, we have

$$f_{n+1}^2 - f_n^2 = \partial(f_n^2)^*(D_t\sigma(n) - D_t\alpha(n))\Delta t + f^2(D_t\sigma(n) - D_t\alpha(n))\Delta t^2.$$

Since the first term of the right side vanishes, we have $f_{n+1} \geq f_n$. The situation is the same in the elastic-plastic case . This proves (3).

To prove (4), we note the identity

$$\partial f' = \frac{1}{f'} S(\sigma' - \alpha').$$

Hence we have

$$(\sigma - \alpha)^*\partial f' = \frac{1}{f'}(\sigma - \alpha)^*S(\sigma' - \alpha') = \frac{1}{f'}(\sigma' - \alpha')^*S(\sigma - \alpha)$$

$$= \frac{f}{f'}(\sigma' - \alpha')^*\partial f,$$

which proves (4).

Now we introduce the quantity

$$H(n) = \|D_t u(n)\|_\rho^2 + \frac{1}{\eta}(S\alpha(n),\alpha(n+1)) + (C\sigma(n), \sigma(n+1)),$$

where $\|u\|_\rho^2$ denotes $\sum (\rho u_i, u_i)$.

THEOREM 7.6. For any n (≥ 1) it holds that

$$(3.7) \qquad D_{\bar{t}}H(n) \leq 0.$$

PROOF. From the discrete equation of motion, it is easy to get the following identity :

$$D_{\bar{t}}\|D_t u(n)\|_\rho^2 + (\sigma(n),(D_t + D_{\bar{t}})\varepsilon(n)) = 0.$$

Taking (5) of Theorem 7.5 into account, we have

$$0 = -(\sigma(n),(D_t + D_{\bar{t}})\varepsilon(n)) + \frac{1}{\eta}(\sigma(n),S(D_t + D_{\bar{t}})\alpha(n)) + (\sigma(n),C(D_t + D_{\bar{t}})\sigma(n))$$

$$= D_{\bar{t}}H(n) + \frac{1}{\eta}(\sigma(n) - \alpha(n),S(D_t + D_{\bar{t}})\alpha(n)).$$

In what follows we show that the second term of the right side of the last equality in nonnegative. Define $(u,v)_e = \int_e u^* v dx$ for an arbitrary element e and put

$$Q_e = (\sigma(n) - \alpha(n),S(D_t + D_{\bar{t}})\alpha(n))_e.$$

We shall check all possible cases to show $Q_e \geq 0$.

(1). Both $\sigma(n)$ and $\sigma(n+1)$ are determined by the elastic rule. In this case, clearly $Q_e = 0$.

(2). Both $\sigma(n)$ and $\sigma(n+1)$ are determined by the plastic rule. In this case, we have by (1) of Theorem 7.5

$$(\sigma(n) - \alpha(n), \partial f_{n-1})_e = (\sigma(n-1) - \alpha(n-1), \partial f_{n-1})_e \geq 0.$$

Therefore

$$Q_e = (\sigma(n) - \alpha(n), \partial f_n)_e \partial f_n^* D_t \sigma(n)$$

$$+ (\sigma(n) - \alpha(n), \partial f_{n-1})_e \partial f_{n-1}^* D_t \sigma(n-1) \geq 0.$$

(3). $\sigma(n)$ is determined by the elastic (or plastic) rule and $\sigma(n+1)$ by the plastic (or elastic) rule. This case is the same as case (2).

(4). $\sigma(n)$ is determined by the elastic rule and $\sigma(n+1)$ by the elastic-plastic rule. By (4) of Theorem 7.5 we have

$$(\sigma(n) - \alpha(n), \partial \bar{f}_n)_e = k_n(\bar{\sigma}(n) - \alpha(n), \partial f_n)_e$$

$$= k_n(\sigma(n) - \alpha(n), \partial f_n)_e + k_n(\bar{\sigma}(n) - \sigma(n), \partial f_n)_e \geq 0.$$

Therefore we have

$$Q_e = (\sigma(n) - \alpha(n), \partial \bar{f}_n)_e \frac{\partial \bar{f}_n(\sigma(n+1) - \bar{\sigma}(n))}{\Delta t} \geq 0.$$

(5). The case where $\sigma(n)$ and $\sigma(n+1)$ are determined by the plastic and elastic-plastic rules, respectively, does not occur.

(6). $\sigma(n)$ is determined by the elastic-plastic rule and $\sigma(n+1)$ by the plastic rule. In this case, we have the identity

$$(\sigma(n) - \alpha(n), \partial \bar{f}_{n-1})_e = (\bar{\sigma}(n-1) - \alpha(n-1), \partial \bar{f}_{n-1})_e$$

$$+ (\sigma(n) - \bar{\sigma}(n-1) - [\alpha(n) - \alpha(n-1)], \partial \bar{f}_{n-1})_e.$$

This quantity is nonnegative, since the first term of the right side is nonnegative and the second one vanishes by (2) of Theorem 7.5. Hence we have

$$Q_e = (\ \sigma(n) - \alpha(n), \partial\bar{f}_{n-1})_e \ \frac{\partial\bar{f}_{n-1}(\sigma(n) - \bar{\sigma}(n-1))}{\Delta t}$$

$$+ (\ \sigma(n) - \alpha(n), \partial f_n)_e \partial f_n^* D_t \sigma(n) \geq 0.$$

(7). The case where $\sigma(n)$ and $\sigma(n+1)$ are determined by the elastic-plastic and elastic rules, respectively, is now evident. This completes the proof.

We want to show that H(n) is positive under a certain condition. To do this we introduce E(n) and R(n) defined by

$$E(n) = \|D_t u(n)\|_\rho^2 + \frac{1}{2\eta}(\|\alpha(n)\|_S^2 + \|\alpha(n+1)\|_S^2) + \frac{1}{2}(\|\sigma(n)\|_C^2 + \|\sigma(n+1)\|_C^2),$$

$$R(n) = \frac{1}{2\eta}\|D_t\alpha(n)\|_S^2 \Delta t^2 + \frac{1}{2}\|D_t\sigma(n)\|_C^2 \Delta t^2 ,$$

where $\|\sigma\|_C^2 = (C\sigma,\sigma)$ and $\|\alpha\|_S^2 = (S\alpha\ ,\alpha)$. Clearly $H(n) = E(n) - R(n)$.

We first prove that

$$R(n) \leq \frac{\Delta t^2}{2} (DD_t\varepsilon(n), D_t\varepsilon(n)). \tag{3.8}$$

Introduce the quantity

$$\bar{\bar{D}}'(n) = (1 - \theta_n)\bar{D}'(n)$$

to represent the stress-strain and α-strain relations in the form

$$D_t\sigma(n) = (D - \bar{\bar{D}}'(n))D_t\varepsilon(n),$$

$$D_t\alpha(n) = \eta S^{-1}C\bar{\bar{D}}'(n)D_t\varepsilon(n),$$

with a suitable choice of θ_n. R(n) is then written as

$$\frac{2}{\Delta t^2} R(n) = \eta \|S^{-1} C\bar{\bar{D}}'(n) D_t \varepsilon(n)\|_S^2 + \|(D - \bar{\bar{D}}'(n)) D_t \varepsilon(n)\|_C^2 .$$

Here we put

$$S(n) = \eta\|S^{-1}C\bar{\bar{D}}'(n)D_t\varepsilon(n)\|_S^2 - 2(\bar{\bar{D}}'(n)D_t\varepsilon(n), D_t\varepsilon(n))$$

$$+ (\bar{\bar{D}}'(n)C\bar{\bar{D}}'(n)D_t\varepsilon(n), D_t\varepsilon(n))$$

$$= ([\eta\bar{\bar{D}}'(n)CS^{-1}C\bar{\bar{D}}'(n) - 2\bar{\bar{D}}'(n) + \bar{\bar{D}}'(n)C\bar{\bar{D}}'(n)]D_t\varepsilon(n), D_t\varepsilon(n)),$$

so that

$$\frac{2}{\Delta t^2} R(n) = (DD_t\varepsilon(n), D_t\varepsilon(n)) + S(n).$$

$S(n)$ is nonpositive. To prove this, put $A_n = \eta + \partial\bar{f}_n^* D \partial\bar{f}_n$. Then, since

$$\text{(3.9)} \qquad \bar{\bar{D}}'(n) = (1 - \theta_n) \frac{\partial\bar{f}_n \partial\bar{f}_n^* D}{A_n}$$

and $\partial\bar{f}_n^* S^{-1} \partial\bar{f}_n = \partial\bar{f}_n^*(\bar{\sigma}(n) - \alpha(n))/\bar{f}_n = 1$, we have

$$\bar{\bar{D}}'(n)CS^{-1}C\bar{\bar{D}}'(n) = (1 - \theta_n) \frac{\bar{\bar{D}}'(n)}{A_n} .$$

Therefore $S(n)$ can also be written as

$$S(n) = ([\eta(1 - \theta_n)/A_n - 1]\, \bar{\bar{D}}'(n)D_t\varepsilon(n), D_t\varepsilon(n))$$

$$+ ([\bar{\bar{D}}'(n)C\bar{\bar{D}}'(n) - \bar{\bar{D}}'(n)]D_t\varepsilon(n), D_t\varepsilon(n)).$$

Now, since $\bar{\bar{D}}'(n)$ is nonnegative definite, the first term of the right side is nonpositive. This is the case for the second term too, since substituting (3.9) into this term it can be written

$$\frac{1-\theta_n}{A_n} ([1 - \theta_n]\partial\bar{f}_n^* D \partial\bar{f}_n/A_n - 1)\, [\partial\bar{f}_n^* DD_t\varepsilon(n)]^2 \le 0.$$

Hence the nonpositivity of $S(n)$ is proved and the inequality (3.8) holds well.

THEOREM 7.7. Let h be the maximum length of all the sides of the triangles of Ω. There are positive constants ξ and k such that if

$$\rho - \xi \frac{\Delta t^2}{h^2} > 0, \tag{3.10}$$

then for any n ($\geq$ 1) the following inequality holds.

$$k\|D_t u(n)\|_\rho^2 + \frac{1}{2\eta}(\|\alpha(n)\|_S^2 + \|\alpha(n+1)\|_S^2)$$
$$+ \frac{1}{2}(\|\sigma(n)\|_C^2 + \|\sigma(n+1)\|_C^2) \leq \|D_t u(0)\|_\rho^2.$$

PROOF. It is known that there is a positive constant ξ depending only on the regularity of the triangulation of Ω such that

$$\frac{\Delta t^2}{2}(DD_t\varepsilon(n), D_t\varepsilon(n)) \leq \frac{\xi}{\rho}\frac{\Delta t^2}{h^2}\| D_t u(n)\|_\rho^2.$$

Therefore we have

$$\|D_t u(n)\|_\rho^2 - R(n) \geq (1 - \xi\Delta t^2/\rho h^2)\|D_t u(n)\|_\rho^2.$$

Set $k = 1 - \xi\Delta t^2/\rho h^2$ and use (3.7) to get the desired inequality.

As is seen in the above proof, the plasticity does not influence the stability criterion. In fact, as proved in [6], the stability of the finite element scheme to approximate elastic vibration problems is ensured by condition (3.10) if the basis function is the same. The constant ξ is estimated in [6].

7.4. Isotropic hardening problem

The main result obtained for the kinematic hardening problem is valid for the isotropic case. Let us seek the approximate solution at time step n in the form

$$u_i(n) = \sum_{p\in P} u_i^p(n)\phi_p$$

Also, the equation of motion is approximated as

$$(\rho\, D_t D_{\bar t} u_i(n), \phi_p) + \sum_j (\sigma_{ij}(n), \phi_{p,j}) = 0 \qquad p \in P.$$

We define the elastic, plastic and elastic-plastic rules as follows :

DEFINITION. (1). ELASTIC RULE. $\sigma(n)$ lies in or on the yield surface of step(n) and

$(4.1)_a$ $\quad D_t\sigma(n) = DD_t\varepsilon(n)$.

$(4.1)_b$ The yield surface of step(n+1) is

$$f(\tau) = \max_{m\le n+1} (H(0), f(\sigma_m)).$$

(2). PLASTIC RULE. $\sigma(n)$ lies on the yield surface of step(n) and

$(4.2)_a$ $$D_t\sigma(n) = (D - D'_n)D_t\varepsilon(n), \quad D'_n = \frac{D\partial f_n \partial f_n^* D}{H'(H^{-1}(f_n)) + \partial f^* D \partial f_n}$$

$(4.2)_b$ The yield surface of step(n+1) is

$$f(\tau) = f(\sigma(n+1)).$$

(3). ELASTIC-PLASTIC RULE. $\sigma(n)$ lies in the yield surface of step(n) and the point $\tilde\sigma(n+1)$ determined by the elastic rule from $\sigma(n)$ is outside the yield surface of step(n) and $\partial f_n^*(\tilde\sigma(n+1) - \sigma(n)) \ge 0$ holds. Then we choose θ_n so that the point $\bar\sigma(n) = \sigma(n) + \theta_n(\tilde\sigma(n+1) - \sigma(n))$ comes on the yield surface of step(n), and we define $\bar f_n$, $\partial\bar f_n$ and $\bar\varepsilon(n)$ as in the kinematic hardening case. Then

$(4.3)_a$ $\sigma(n+1) - \bar{\sigma}(n) = (D - \bar{D}'_n)(\varepsilon(n+1) - \bar{\varepsilon}(n))$, $\bar{D}'_n = D'_n(\ \sigma(n) + \bar{\sigma}(n)\ )$.

$(4.3)_b$ The yield surface of step(n+1) is

$$f(\tau) = f(\sigma(n+1)).$$

REMARK. When the elastic-plastic rule is applied, the vector $\sigma(n+1) - \bar{\sigma}(n)$ is again transversal to the yield surface at the stress point $\bar{\sigma}(n)$ as in the kinematic hardening case.

The discrete stress-strain relation in the isotropic hardening problem is defined in the same way as before. For the sake of simplicity, we start from $u(0) = \sigma(0) = 0$.

(A). Determine $\tilde{\sigma}(n+1)$ $(n \geq 0)$ by the elastic rule if $\sigma(n)$ lies in the yield surface of step(n). If $\tilde{\sigma}(n+1)$ is still in it we define $\sigma(n+1) = \tilde{\sigma}(n+1)$.

(B). If $\sigma(n)$ is in the yield surface of step(n), and if $\tilde{\sigma}(n+1)$ determined by the elastic rule comes on or outside the yield surface of step(n), then $\sigma(n+1)$ is determined by

$(B)_a$ the elastic-plastic rule if $\partial f^*_n(\ \tilde{\sigma}(n+1) - \sigma(n)) \geq 0$, and otherwise by

$(B)_b$ the elastic rule.

The subsequent relations are given as follows :

(C). If $\sigma(n+1)$ is on the yield surface of step(n+1), then determine $\tilde{\sigma}(n+2)$ by the plastic rule, and define

$(C)_a$ $\sigma(n+2) = \tilde{\sigma}(n+2)$ if $\partial f^*_{n+1}\ (\ \tilde{\sigma}(n+2) - \sigma(n+1)) \geq 0$, and otherwise

$(C)_b$ determine $\sigma(n+2)$ anew by the elastic rule.

(D). If $\sigma(n+1)$ remains in the yield surface of step(n+1), then return to procedure (A) ⟶ (B), replacing n by n+1.

REMARK. For $\tilde{\sigma}(n+2)$ determined by the elastic rule in case (C) assume

$$\partial f^*_{n+1}(\tilde{\sigma}(n+2) - \sigma(n+1)) < 0.$$

Then $\sigma(n+2)$ is determined anew by the elastic rule. In this case the relation

$$\partial f^*_{n+1}(\sigma(n+2) - \sigma(n+1)) < 0$$

holds again. The reason is the same in the kinematic hardening case.

The plasticity has no influence on the stability of this explicit integration scheme. In fact, we have

THEOREM 7.8. Under the same condition of the preceding theorem we have the following a priori estimate for the discrete solution :

$$k\|D_t u(n)\|^2_\rho + \frac{1}{2}(\|\sigma(n)\|^2_C + \|\sigma(n+1)\|^2_C) \leq \|D_t u(0)\|^2_\rho \qquad (k > 0).$$

PROOF. We define the discrete plastic strain $\varepsilon^p(n)$ by

$$D_t\varepsilon^p(n) = D_t\varepsilon(n) - CD_t\sigma(n).$$

There are three cases, depending on the rule for determining $\sigma(n+1)$:

$$D_t\varepsilon^p(n) = \begin{cases} 0 & \text{elastic rule} \\ \partial f_n \; \partial f^*_n D_t\sigma(n)/H'_n & \text{plastic rule} \\ \partial\bar{f}_n \; \partial\bar{f}^*_n(\sigma(n+1) - \bar{\sigma}(n))/(\Delta t\bar{H}'_n) & \text{elastic-plastic rule} \end{cases}$$

where $H'_n = H'(H^{-1}(f_n))$, $\bar{H}'_n = H'(f_n \to \bar{f}_n)$.

Multiplying both sides of the discrete equation of motion by $(D_t + D_{\bar{t}})u_n$, we have

(4.4) $$D_{\bar{t}}[\|D_t u(n)\|_\rho^2 + (C\sigma(n),\sigma(n+1))] + \sum_e Q_e = 0,$$

where Q_e is the integration of $\sigma(n)^*(D_t + D_{\bar{t}})\varepsilon^P(n)$ over the element e. We shall show below that $Q_e \geq 0$. If this is the case, then the theorem can be proved in the same way as Theorem 7.7.

(1). Both $\sigma(n)$ and $\sigma(n+1)$ are determined by the elastic rule. In this case, it clearly holds that $D_t\varepsilon^P(n) = D_t\varepsilon^P(n-1) = 0$, so that $Q_e = 0$.

(2). Both $\sigma(n)$ and $\sigma(n+1)$ are determined by the plastic rule. Clearly $(\sigma(n), D_t\varepsilon^P(n))_e \geq 0$. Since $\partial f^*_{n-1}(\sigma(n) - \sigma(n-1)) \geq 0$, we have

$$\partial f^*_{n-1}\,\sigma(n) \geq \partial f^*_{n-1}\,\sigma(n-1) = f_{n-1} \geq 0,$$

from which follows $Q_e \geq 0$.

(3). $\sigma(n)$ is determined by the elastic rule and $\sigma(n+1)$ by the plastic rule. In this case $Q_e \geq 0$ is clear, since $D_t\varepsilon^P(n-1) = 0$.

(4). $\sigma(n)$ is determined by the plastic rule and $\sigma(n+1)$ by the elastic rule. This case is the same as case (2).

(5). $\sigma(n)$ is determined by the elastic rule and $\sigma(n+1)$ by the elastic-plastic rule. In this case, it holds that $D_{\bar{t}}\varepsilon^P(n) = 0$. Hence

$$Q_e = (\sigma(n), D_t\varepsilon^P(n)) = (\sigma(n), \frac{1}{H'_n}\partial\bar{f}_n\,\partial\bar{f}^*_n(\sigma(n+1) - \bar{\sigma}(n)))/\Delta t.$$

Set $k_n = f_n/\bar{f}_n$. Then, since $f_n \partial f_n = S\sigma(n)$, we have

$$(\partial\bar{f}_n, \sigma(n))_e = 1/\bar{f}_n\,(S\bar{\sigma}(n), \sigma(n))_e = k_n(\partial f_n, \bar{\sigma}(n))_n$$

$$= k_n(\partial f_n, \sigma(n) + \bar{\sigma}(n) - \sigma(n))_e \geq 0,$$

so that $Q_e \geq 0$.

(6). $\sigma(n)$ is determined by the elastic-plastic rule and $\sigma(n+1)$ by the plastic rule. First, we see $(\sigma(n), D_t \varepsilon^p(n))_e \geq 0$. On the other hand,

$$(\sigma(n), D_{\bar{t}} \varepsilon^p(n))_e = \frac{1}{\bar{H}'_{n-1}} (\sigma(n), \partial\bar{f}_{n-1})_e \partial\bar{f}^*_{n-1}(\sigma(n) - \bar{\sigma}(n-1))/\Delta t.$$

Since $\partial\bar{f}^*_{n-1}\sigma(n) = \partial\bar{f}^*_{n-1}(\bar{\sigma}(n-1) + \sigma(n) - \bar{\sigma}(n-1)) \geq 0$, we have $Q_e \geq 0$.

$Q_e \geq 0$ is obvious in the last case (elastic-plastic and elastic). Thus it is nonnegative in any situation, and we have by (4.4)

$$D_{\bar{t}}[\|D_t u(n)\|^2_\rho + (C\sigma(n), \sigma(n+1))] \leq 0,$$

from which follows the theorem.

CHAPTER 8

EXPLICIT SCHEMES FOR QUASI-STATIC PROBLEMS

8.1. An explicit method for the system with multiple degrees of freedom

In this chapter we analyze some explicit approximating methods to solve the quasi-static problems. To explain some essential features of the methods and the analysis, we first consider an approximate method for the multiple system introduced in Chapter 4 :

(1.1) $$\sigma_i - \sigma_{i+1} = b_i(t) \qquad i = 1 \sim N$$

(1.2) $\dot{\sigma}_i = k_i\dot{U}_i$, $\dot{\alpha}_i = 0$ if $|\sigma_i - \alpha_i| < z_0$, or $|\sigma_i - \alpha_i| = z_0$ and $(\sigma_i - \alpha_i)\dot{U}_i < 0$

(1.3) $\dot{\sigma}_i = (1 - \xi_i)k_i\dot{U}_i$, $\dot{\alpha}_i = \dot{\sigma}_i$ if $|\sigma_i - \alpha_i| = z_0$ and $(\sigma_i - \alpha_i)\dot{U}_i \geq 0$,

where $U_i = u_i - u_{i-1}$ ($u_0 = u_{N+1} = 0$). The original problem is to seek (u, σ, α) satisfying these equations in $I = (0,T)$ with the initial conditions $u_i = \sigma_i = \alpha_i = 0$. In this chapter the system (1.1) is expressed simply as

$$\Delta\sigma = b(t).$$

Let E be the set of integers $\{1,2,\ldots,N+1\}$. We discretize the problem (1.1)~(1.3) according to the following rule. All of the springs are elastic until $t = t_0$ at which some spring satisfies

$$| \sigma_i^n - \alpha_i^n | = z_0,$$

where (u^n,σ^n,α^n) is the solution of the problem

$$\Delta\sigma^n = b(t_n) \qquad \sigma_i^n = k_i U_i^n, \quad \alpha_i^n = 0.$$

Let $0 < t_0 < t_1 < \quad < t_M = T$ be a partition of the time interval I. Although the time mesh need not be equal (as will be seen in the following discussion), we set $\Delta t = t_{n+1} - t_n$ = constant for simplicity's sake.
We define the set

$$S_i^n = \{ \tau \; ; \; | \tau - \alpha_i^n | = z_0 \}$$

for each $i \in E$, and call it the yield surface of spring i at stage n. In reality, this surface consists of two points for the present problem.

The computation of $(u^{n+1}, \sigma^{n+1}, \alpha^{n+1})$ takes 8 steps.

Step 1. Classify E as $E = P^n + E^n$, where

$$P^n = \{ i \in E \; ; \; |\sigma_i^n - \alpha_i^n| = z_0 \}$$

$$E^n = \{ i \in E \; ; \; |\sigma_i^n - \alpha_i^n| < z_0 \},$$

and set k = 1.

Step 2. Compute $(u^{n+1,k}, \sigma^{n+1,k})$ by solving the equation

(1.4) $$\Delta\sigma^{n+1} = b(t_{n+1})$$

under the relation

$$D_t\sigma_i^n = (1 - \xi_i) k_i D_t U_i^n \qquad \text{in } D_+ = \{u^{n+1} \in R^N \; ; \; (\sigma_i^n - \alpha_i^n) D_t U_i^n \geq 0 \}$$
$$D_t\sigma_i^n = k_i D_t U_i^n \qquad \text{in } D_- = \{u^{n+1} \in R^N \; ; \; (\sigma_i^n - \alpha_i^n) D_t U_i^n < 0 \}$$

to P^n, and to E^n:

$$D_t\sigma_i^n = k_i D_t U_i^n \; .$$

Step 3. Classify E^n as $E^n = E_0^n + (\text{new})E^n$, where

$$E_0^n = \{\, i \in E^n \,;\, |\sigma_i^{n+1,k} - \alpha_i^n| \geq z_0 \}$$
$$(\text{new})\, E^n = \{\, i \in E^n \,;\, |\sigma_i^{n+1,k} - \alpha_i^n| < z_0 \}$$

Set $E_P^n = \Phi$ ..... empty.

Step 4. For $i \in E_0^n$, determine such $\{\bar{\sigma}_i^n, \bar{U}_i^n\} \in R^2$ that

$$\bar{\sigma}_i^n = \sigma_i^n + \theta(\sigma_i^{n+1,k} - \sigma_i^n) \qquad (0< \theta \leq 1) \quad \text{and} \quad \bar{\sigma}_i^n \in S_i^n,$$

$$\bar{U}_i^n - U_i^n = \frac{1}{k_i}(\bar{\sigma}_i^n - \sigma_i^n).$$

Set $E_P^n = E_P^n + E_0^n$.

Step 5. Compute $(u^{n+1,k+1}, \sigma^{n+1,k+1})$ by solving

(1.5) $$\Delta\sigma^{n+1} = b(t_{n+1})$$

under the relation

(A) $$\begin{cases} D_t\sigma_i^n = (1 - \xi_i)k_i D_t U_i^n & \text{in } D_+ = \{u^{n+1} \in R^N \;;\; (\sigma_i^n - \alpha_i^n)D_t U_i^n \geq 0\} \\ D_t\sigma_i^n = k_i D_t U_i^n & \text{in } D_- = \{u^{n+1} \in R^N \;;\; (\sigma_i^n - \alpha_i^n)D_t U_i^n < 0\} \end{cases}$$

to P^n, and

(B) $$\begin{cases} \bar{D}_t\sigma_i^n = (1 - \xi_i)k_i \bar{D}_t U_i^n & \text{in } D_+ = \{u^{n+1} \in R^N \;;\; (\bar{\sigma}_i^n - \alpha_i^n)\bar{D}_t U_i^n \geq 0\} \\ \bar{D}_t\sigma_i^n = k_i \bar{D}_t U_i^n & \text{in } D_- = \{u^{n+1} \in R^N \;;\; (\bar{\sigma}_i^n - \alpha_i^n)\bar{D}_t U_i^n < 0\} \end{cases}$$

to E_P^n, where $\bar{D}_t\sigma_i^n = (\sigma_i^{n+1} - \bar{\sigma}_i^n)/\Delta t$ and $\bar{D}_t U_i^n = (U_i^{n+1} - \bar{U}_i^n)/\Delta t$, and

(C) $$D_t\sigma_i^n = k_i D_t U_i^n$$

to E^n.

Step 6. Define the new E_0^n by

$$E_0^n = \{ i \in E^n ; | \sigma_i^{n+1,k+1} - \alpha_i^n | \geq z_0 \}.$$

Step 7. If E_0^n is empty, then go to step 8, else define new E^n by $E^n = E^n - E_0^n$ and set $k = k+1$, return to step 4.

Step 8. Define $(u^{n+1}, \sigma^{n+1}) = (u^{n+1,k+1}, \sigma^{n+1,k+1})$. Determine α_i^{n+1} by :

$$D_t \alpha_i^n = 0 \qquad \text{for } E^n$$

$$\left\{\begin{array}{ll} D_t\alpha_i^n = D_t\sigma_i^n & \text{if } (\sigma_i^n - \alpha_i^n) D_t U_i^n \geq 0 \\ D_t \alpha_i^n = 0 & \text{if } (\sigma_i^n - \alpha_i^n) D_t U_i^n < 0 \end{array}\right. \qquad \text{for } P^n$$

$$\left\{\begin{array}{ll} D_t\alpha_i^n = \bar{D}_t\sigma_i^n & \text{if } (\bar{\sigma}_i^n - \alpha_i^n) \bar{D}_t U_i^n \geq 0 \\ D_t \alpha_i^n = 0 & \text{if } (\bar{\sigma}_i^n - \alpha_i^n) \bar{D}_t U_i^n < 0 \end{array}\right. \qquad \text{for } E_p^n .$$

This procedure is well defined. In fact, it is clear that the iteration [step 4 → step 7] terminates with finite k. Also, each of the problems (1.4) and (1.5) has a unique solution. We shall prove below that (1.5) has a unique solution since (1.5) is more general than (1.4). To do this we show that (1.5) has an equivalent variational problem. Consider the functional

$$(1.6) \qquad F(u^{n+1}) = \frac{1}{2}(D_t\sigma^n, D_t U^n) + \frac{1}{2} \sum_i \xi_i k_i (\bar{U}_i^n - U_i^n)/\Delta t \cdot \lambda_i - (D_t b^n, D_t u^n)$$

with the conditions (A),(B) and (C), where

$$\lambda_i = \left\{\begin{array}{lll} 0 & & \text{for } i \in E^n + P^n \\ D_t U_i^n & \text{in } D_+ & \multirow{2}{*}{for $i \in E_p^n$.} \\ (\bar{U}_i^n - U_i^n)/\Delta t & \text{in } D- & \end{array}\right.$$

THEOREM 8.1. There is a unique minimum point of $F(u^{n+1})$ which is the solution of (1.5).

PROOF. For each $i \in E$, set

$$F_i(u^{n+1}) = \frac{1}{2} D_t\sigma_i^n \, D_t U_i^n + \frac{1}{2}\xi_i k_i(\bar{U}_i^n - U_i^n)/\Delta t \cdot \lambda_i - D_t b_i^n \, D_t u_i^n .$$

(1). F_i is continuous as a function of u^{n+1}. This is clear for $i \in E^n + P^n$. For $i \in E_p^n$, it holds that

$$D_t\sigma_i^n = (1 - \xi_i)k_i\bar{D}_t U_i^n + k_i(\bar{U}_i^n - U_i^n)/\Delta t \tag{1.7}$$

$$= k_i D_t U_i^n - \xi_i k_i \bar{D}_t U_i^n$$

in D_+, and that

$$D_t\sigma_i^n = k_i D_t U_i^n \tag{1.8}$$

in D_-. Hence $D_t\sigma_i^n$ is continuous across the surface between D_+ and D_-. Since λ_i is continuous, F_i is too.

(2). F_i is continuously differentiable on u^{n+1}. This is clear for $i \in E^n$. If $i \in P^n$, let v be an arbitrary component of u^{n+1}. Since $D_t\sigma_i^n = (1 - \xi_i)k_i D_t U_i^n$ (or $= k_i D_t U_i^n$) in D_+ (or in D_-), we have in D_+

$$\partial F_i/\partial v = (1 - \xi_i)k_i D_t U_i^n \, \partial D_t U_i^n/\partial v - D_t b_i^n \, \partial D_t u_i^n/\partial v$$

and in D_-

$$\partial F_i/\partial v = k_i D_t U_i^n \, \partial D_t U_i^n/\partial v - D_t b_i^n \, \partial D_t u_i^n/\partial v.$$

Since $D_t U_i^n = 0$ on the surface between D_+ and D_-, these derivatives coincide on this surface. On the other hand, if $i \in E_p^n$, then by (1.7) and (1.8) we have in D_+

$$\partial F_i/\partial v = \frac{\partial}{\partial v}\left[\frac{1}{2}(k_i D_t U_i^n - \xi_i k_i \bar{D}_t U_i^n) D_t U_i^n + \frac{1}{2}\xi_i k_i(\bar{U}_i^n - U_i^n)/\Delta t \cdot D_t U_i^n - D_t b_i^n D_t u_i^n\right]$$

$$= (1 - \xi_i)k_i D_t U_i^n \, \partial D_t U_i^n/\partial v + \xi_i k_i(\bar{U}_i^n - U_i^n)/\Delta t \cdot \partial D_t U_i^n/\partial v - D_t b_i^n \partial D_t u_i^n/\partial v$$

$$= (k_i D_t U_i^n - \xi_i k_i \bar{D}_t U_i^n)\partial D_t U_i^n/\partial v - D_t b_i^n \ \partial D_t u_i^n/\partial v,$$

and in D_-

$$\partial F_i/\partial v = k_i D_t U_i^n \ \partial D_t U_i^n/\partial v - D_t b_i^n \partial D_t u_i^n/\partial v,$$

so that $\partial F_i/\partial v$ is continuous in the whole u^{n+1}- space.

(3). $F(u^{n+1})$ is a (piecewise) quadratic form which is bounded below. Hence it has a minimum point which is also stationary by (1) and (2) above. The stationary condition $\partial F_i/\partial v = 0$ is written as

$$\sum_{i=1}^{N} (D_t\sigma_i^n \frac{\partial D_t U_i^n}{\partial v} - D_t b_i^n \frac{\partial D_t u_i^n}{\partial v}) = 0 \ ;$$

that is, $\Delta D_t \sigma^n = D_t b^n$. Hence the pair of the minimum point u^{n+1} and σ^{n+1} which is determined by the conditions (A),(B) and (C),is a solution of (1.5).

(4). The uniqueness of the solution of (1.5) follows from the uniqueness of the minimum point of $F(u^{n+1})$, which is proved in the same way as in the semidiscrete problem. Thus the theorem is proved.

REMARK. To solve (1.4) or (1.5) we apply the method of trial and error assuming a suitable stress-strain relation, as is usually done. If b(t) has good properties, then no spring will change the state so frequently, and the method of trial and error will work effectively.

8.2. The order of convergence with respect to Δt

To estimate the speed of convergence of the approximate solution, we prepare two theorems. In this section we assume that $k_i = k$, $\xi_i = \xi$, for simplicity's sake.

THEOREM 8.2. Let (u^n,σ^n,α^n) $(n = 1 \sim M)$ be the approximate solution obtained by the above method. Then the following relations hold for any n.

$$(2.1) \qquad \sum_i (D_tU_i^n - \frac{1}{k} D_t\sigma_i^n)(\tau_i - \sigma_i^{n+1}) \le 0 \qquad \text{for all } \tau_i\text{: } |\tau_i - \alpha_i^n| \le z_0$$

$$(2.2) \qquad D_t\alpha_i^n = \frac{(1-\xi)k}{\xi} (D_tU_i^n - \frac{1}{k} D_t\sigma_i^n) \qquad \text{for all } i \in E.$$

PROOF. At step 8 of each stage of integration, we classify the springs as

$$E = E^n + P^{n,P} + P^{n,E} + E_P^{n,P} + E_P^{n,E}$$

where

$$P^{n,P} = \{i \in P^n \; ; \; (\sigma_i^n - \alpha_i^n)D_tU_i^n \ge 0 \}$$
$$P^{n,E} = \{i \in P^n \; ; \; (\sigma_i^n - \alpha_i^n)D_tU_i^n < 0 \}$$
$$E_P^{n,P} = \{i \in E_P^n \; ; \; (\bar{\sigma}_i^n - \alpha_i^n)\bar{D}_tU_i^n \ge 0 \}$$
$$E_P^{n,E} = \{i \in E_P^n \; ; \; (\bar{\sigma}_i^n - \alpha_i^n)\bar{D}_tU_i^n < 0 \}.$$

Now since

$$D_tU_i^n - \frac{1}{k}D_t\sigma_i^n = \begin{cases} 0 & i \in E^n + P^{n,E} + E_P^{n,E} \\ \xi D_tU_i^n & i \in P^{n,P} \\ \xi\bar{D}_tU_i^n & i \in E_P^{n,P}, \end{cases}$$

we have

$$(2.3) \qquad (D_tU_i^n - \frac{1}{k}D_t\sigma_i^n)(\tau_i - [\sigma_i^n]) \le 0 \qquad \text{for all } \tau_i : |\tau_i - \alpha_i^n| \le z_0$$

where $[\sigma_i^n] = \sigma_i^n$ for $i \in E - E_P^n$ and $= \bar{\sigma}_i^n$ for $i \in E_P^n$. Since the sign of $\sigma_i^{n+1} - [\sigma_i^n]$ for $i \in P^{n,P} + E_P^{n,P}$ is the same as that of $[\sigma_i^n] - \alpha_i^n$, we have

$$0 \ge (D_tU_i^n - \frac{1}{k}D_t\sigma_i^n)((\tau_i - \sigma_i^{n+1}) + (\sigma_i^{n+1} - [\sigma_i^n]))$$

$$\geq (D_t U_i^n - \frac{1}{k} D_t \sigma_i^n)(\tau_i - \sigma_i^{n+1}),$$

which proves (2.1). The identity (2.2) is obvious for $i \in E^n + P^{n,E} + E_P^{n,E}$, since $D_t \alpha_i^n = D_t U_i^n - D_t \sigma_i^n / k = 0$ for these points. For $i \in P^{n,P}$ and $i \in E_P^{n,P}$, respectively, it holds that

$$\begin{cases} D_t U_i^n - \frac{1}{k} D_t \sigma_i^n = \xi D_t U_i^n \\ D_t \alpha_i^n = D_t \sigma_i^n, \end{cases} \qquad \begin{cases} D_t U_i^n - \frac{1}{k} D_t \sigma_i^n = \xi \bar{D}_t U_i^n \\ D_t \alpha_i^n = \bar{D}_t \sigma_i^n, \end{cases}$$

from which (2.2) follows. This completes the proof.

Before estimating the error of the approximate solution, we provide

THEOREM 8.3. Let (u,σ,α) and (u^n,σ^n,α^n) be the solution of (1.1) ~ (1.3) and the approximate solution obtained by the present method, respectively. Then the following a priori estimates hold.

$$\|\dot{U}\|_\infty , \|\dot{\sigma}\|_\infty , \|\dot{\alpha}\|_\infty \leq C, \tag{2.4}$$

$$\|D_t U^n\| , \|D_t \sigma^n\| , \|D_t \alpha^n\| \leq C, \tag{2.5}$$

where C is a constant, independent of n and Δt, and $\|U\|^2 = (U,U) = \sum U_i^2$, and $\|U\|_\infty = \operatorname{ess\cdot sup}_I \|U\|$.

PROOF. Estimates (2.4) are already obtained in Theorem 4.6 of Chapter 4. To estimate $D_t \sigma^n$, we set $\tau_i = \sigma^n$ in (2.1) to get

$$\sum_i \frac{1}{k}(D_t \sigma_i^n)^2 \leq \sum_i D_t U_i^n \, D_t \sigma_i^n \leq C\|D_t b^n\| \, \|D_t u^n\|. \tag{2.6}$$

Now $|D_t U_i^n| \leq C|D_t \sigma_i^n|$ is clear for $i \in P^n + E^n + E_P^{n,E}$. For $i \in E_P^{n,P}$, since

$$|\bar{U}_i^n - U_i^n| \leq C|\bar{\sigma}_i^n - \sigma_i^n| \leq C|\sigma_i^{n+1} - \sigma_i^n|,$$

we have, by (1.7)

$$|D_t U_i^n| \le |\bar{D}_t U_i^n| + |\bar{U}_i^n - U_i^n|/\Delta t \le C|D_t \sigma_i^n| .$$

Hence we have $\|D_t u^n\| \le C\|D_t U^n\| \le C\ \|D_t \sigma^n\|$. Therefore (2.6) implies $\|D_t \sigma^n\| \le C\|D_t b^n\|$ for a certain constant C independent of n and Δt. This proves (2.5).

THEOREM 8.4. Let (u, σ, α) and $(u^n, \sigma^n, \alpha^n)$ be those in the preceding theorem. Then there is a constant C independent of n, Δt such that

$$\|U(t_n) - U^n\|,\ \|\sigma(t_n) - \sigma^n\|,\ \|\alpha(t_n) - \alpha^n\| \le C\Delta t^{\frac{1}{2}} \qquad (n \le M).$$

PROOF. Remembering the weak form (1.18)~(1.20) of Chapter 4, we Substitute (1.20) into (1.19) and set

$$\tau = \alpha + \theta_{n+1} z_0 \qquad \theta_{n+1} = \frac{\sigma^{n+1} - \alpha^{n+1}}{z_0}$$

Integrating the resulting inequality from t_n to t_{n+1}, we have

$$0 \ge \int_{t_n}^{t_{n+1}} (\dot{\alpha}, \alpha + \theta_{n+1} z_0 - \sigma)dt$$

$$= -\int_{t_n}^{t_{n+1}} (\alpha, \dot{\alpha} - \dot{\sigma})dt + [\alpha\ , \alpha + \theta_{n+1} z_0 - \sigma\ ]_{t_n}^{t_{n+1}}$$

$$= (D_t\alpha(t_n), \alpha(t_{n+1}) + \theta_{n+1} z_0 - \sigma(t_{n+1}))\Delta t$$

$$+ \int_{t_n}^{t_{n+1}} (\alpha(t_n) - \alpha, \dot{\alpha} - \dot{\sigma})dt$$

$$= (D_t\alpha(t_n), \alpha(t_{n+1}) - \alpha^{n+1} - (\sigma(t_{n+1}) - \sigma^{n+1}))\Delta t$$

$$+ \int_{t_n}^{t_{n+1}} (\alpha(t_n) - \alpha, \dot{\alpha} - \dot{\sigma})dt. \tag{2.7}$$

On the other hand, set

$$\tau = \alpha^n + \frac{\sigma(t_{n+1}) - \alpha(t_{n+1})}{z_0} \; z_0$$

in (2.1) to get

$$0 \geq (D_t\alpha^n, \alpha^n - \alpha(t_{n+1}) + \sigma(t_{n+1}) - \sigma^{n+1})\Delta t$$

$$(2.8) \qquad = (D_t\alpha^n, \alpha^{n+1} - \alpha(t_{n+1}) - (\sigma^{n+1} - \sigma(t_{n+1})))\Delta t$$

$$+ (D_t\alpha^n, \alpha^n - \alpha^{n+1})\Delta t.$$

Adding (2.7) and (2.8) we have

$$0 \geq (D_t(\alpha(t_n) - \alpha^n), \alpha(t_{n+1}) - \alpha^{n+1} - (\sigma(t_{n+1}) - \sigma^{n+1}))\Delta t$$

$$(2.9) \qquad - \|D_t\alpha^n\|^2\Delta t^2 + \int_{t_n}^{t_{n+1}} (\alpha(t_n) - \alpha, \dot{\alpha} - \dot{\sigma})dt.$$

By Theorem 8.2, the sum of the second and third terms of (2.9) are bounded by $C\Delta t^2$. Then summing (2.9) with respect to n, we have

$$\|\alpha(t_{n+1}) - \alpha^{n+1}\|^2 + \|\sigma(t_{n+1}) - \sigma^{n+1}\|^2 \leq C \sum_n \Delta t^2 \leq C\Delta t.$$

This completes the proof of the theorem.

8.3. Two-dimensional problems

We apply the idea of the explicit integration method proposed in the preceding section to two-dimensional problems. We first consider the following problem and estimate the error of the approximate solution :

$$(3.1) \qquad \sum_j (\sigma_{ij}, \phi_{p,j}) = (b_i, \phi_p) \qquad p \in P$$

$$(3.2) \qquad \dot{\sigma} = D\dot{\varepsilon}, \; \dot{\alpha} = 0 \qquad \text{if } f(\sigma-\alpha)<z_0 \text{ or } f(\sigma-\alpha)=z_0 \text{ and } \partial f^*\dot{\sigma}< 0,$$

$$(3.3) \qquad \dot{\sigma} = (D - D')\dot{\varepsilon}, \; \dot{\alpha} = (\sigma - \alpha)\frac{\partial f^*\dot{\sigma}}{f} \qquad \text{if } f(\sigma-\alpha)=z_0 \text{ and } \partial f^*\dot{\sigma} \geq 0,$$

where $u_i(t,x) = \sum_p u_i(t,p)\phi_p(x)$. The analysis proceeds in the similar way as before. But the approximate yield surface in the present problem expands in general, causing a slower speed of convergence, at least on the error estimate with respect to Δt.

All the elements are elastic until $t = t_0$, at which some element satisfies

$$f(\sigma^n - \alpha^n) = z_n = z_0,$$

where $(u^n, \sigma^n, \alpha^n)$ is the solution obtained by using the elastic stress-strain relation $\sigma^n = D\varepsilon^n$, $\alpha^n = 0$. We call the set

$$S_e^n = \{ \tau \in R^3 ; f(\tau - \alpha^n) = z_n \}$$

the yield surface of the element e at stage n. The main part of the procedure to compute $(u^{n+1}, \sigma^{n+1}, \alpha^{n+1})$ consists of 8 steps. We express (3.1) symbolically as

$$(\sigma, \phi) = (b, \phi).$$

Step 1. Classify E as $E = P^n + E^n$, where

$$P^n = \{ e \in E ; f(\sigma^n - \alpha^n) = z_n \},$$
$$E^n = \{ e \in E ; f(\sigma^n - \alpha^n) < z_n \},$$

and set $k = 1$.

Step 2. Compute $(u_k^{n+1}, \sigma_k^{n+1})$ by solving the equation

$$(\sigma_k^{n+1}, \phi) = (b^{n+1}, \phi), \tag{3.4}$$

under the rule

$$\begin{cases} D_t\sigma^n = (D - D_n')D_t\varepsilon^n & \text{in } D_+ = \{ u^{n+1} \; ; \; \partial f_n^* D D_t\varepsilon^n \geq 0 \} \\ D_t\sigma^n = DD_t\varepsilon^n & \text{in } D_- = \{ u^{n+1} \; ; \; \partial f_n^* D D_t\varepsilon^n < 0 \} \end{cases}$$

for P^n, where $D_n' = D\partial f_n \partial f_n^* D/(\eta + \partial f_n^* D \partial f_n)$, and

$$D_t\sigma^n = DD_t\varepsilon^n \qquad \text{for } E^n.$$

Step 3. Classify E^n as $E^n = E_0^n + (\text{new})E^n$, where

$$E_0^n = \{ e \in E^n \; ; \; f(\sigma_k^{n+1} - \alpha^n) \geq z_n \}$$

$$(\text{new})E^n = \{ e \in E^n \; ; \; f(\sigma_k^{n+1} - \alpha^n) < z_n \}.$$

Set $E_p^n = \Phi$ ..... empty.

Step 4. For $e \in E_0^n$, determine $(\bar{\sigma}^n, \bar{\varepsilon}^n) \in R^3 \times R^3$ satisfying $\bar{\sigma}^n \in S_e^n$ and

$$\| \bar{\sigma}^n - \sigma^n \|_{R^3} = \min_{\sigma \in S_e^n} \| \sigma - \sigma^n \|_{R^3}, \quad \bar{\sigma}^n - \sigma^n = D(\bar{\varepsilon}^n - \varepsilon^n).$$

Set $E_p^n = E_p^n + E_0^n$ and $\bar{D}_n' = D\partial\bar{f}_n \partial\bar{f}_n^* D/(\eta + \partial\bar{f}_n^* D \partial\bar{f}_n)$, where $\partial\bar{f}_n = \partial f|_{(\bar{\sigma}^n - \alpha^n)}$

Step 5. Compute $(u_{k+1}^{n+1}, \sigma_{k+1}^{n+1})$ by solving the equation

$$(\sigma^{n+1}, \phi) = (b^{n+1}, \phi) \tag{3.5}$$

applying the rule

$$\text{(A)} \quad \begin{cases} D_t\sigma^n = (D - D_n')D_t\varepsilon^n & \text{in } D_+ = \{u^{n+1} \; ; \; \partial f_n^* D D_t\varepsilon^n \geq 0\} \\ D_t\sigma^n = DD_t\varepsilon^n & \text{in } D_- = \{u^{n+1} \; ; \; \partial f_n^* D D_t\varepsilon^n < 0\} \end{cases}$$

for P^n,

$$\text{(B)} \quad \begin{cases} \bar{D}_t\sigma^n = (D - \bar{D}_n')\bar{D}_t\varepsilon^n & \text{in } D_+ = \{u^{n+1} \; ; \; \partial\bar{f}_n^* D\bar{D}_t\varepsilon^n \geq 0\} \\ \bar{D}_t\sigma^n = D\bar{D}_t\varepsilon^n & \text{in } D_- = \{u^{n+1} \; ; \; \partial\bar{f}_n^* D\bar{D}_t\varepsilon^n < 0\} \end{cases}$$

for E_p^n, where $\bar{D}_t\sigma^n = (\sigma^{n+1} - \bar{\sigma}^n)/\Delta t$ and $\bar{D}_t\varepsilon^n = (\varepsilon^{n+1} - \bar{\varepsilon}^n)/\Delta t$, and

(C) $\quad D_t\sigma^n = DD_t\varepsilon^n \qquad$ for E^n.

Step 6. Define new E_0^n by

$$E_0^n = \{ e \in E^n \; ; \; f(\sigma_{k+1}^{n+1} - \alpha^n) \geq z_n \}.$$

Step 7. If E_0^n is empty, then go to step 8. If not, define new E^n and k by $E^n = E^n - E_0^n$ and $k = k + 1$, and return to step 4.

Step 8. Define $(u^{n+1}, \sigma^{n+1}) = (u_{k+1}^{n+1}, \sigma_{k+1}^{n+1})$. Determine α^{n+1} and the yield surface of the next step in this way :

$$\begin{cases} D_t\alpha^n = 0 \\ f(\tau - \alpha^{n+1}) = z_{n+1} = z_n \end{cases} \qquad \text{for } E^n,$$

$$\begin{cases} D_t\alpha^n = (\sigma^n - \alpha^n)\partial f_n^* D_t\sigma^n / f_n & \text{if } \partial f_n^* D_t\sigma^n \geq 0 \\ D_t\alpha^n = 0 & \text{if } \partial f_n^* D_t\sigma^n < 0 \\ f(\tau - \alpha^{n+1}) = z_{n+1} = \mathrm{Max}(z_n, f(\sigma^{n+1} - \alpha^{n+1})) \end{cases} \qquad \text{for } P^n,$$

$$\begin{cases} D_t\alpha^n = (\bar{\sigma}^n - \alpha^n)\partial \bar{f}_n^* \bar{D}_t\sigma^n / \bar{f}_n & \text{if } \partial \bar{f}_n^* \bar{D}_t\sigma^n \geq 0 \\ D_t\alpha^n = 0 & \text{if } \partial \bar{f}_n^* \bar{D}_t\sigma^n < 0 \\ f(\tau - \alpha^{n+1}) = z_{n+1} = \mathrm{Max}(z_n, f(\sigma^{n+1} - \alpha^{n+1})). \end{cases} \qquad \text{for } E_p^n.$$

This procedure is well defined. In fact, problem (3.5), for example, is equivalent to a minimizing problem. The next theorem is a general version of Theorem 8.1, and proved in the same manner.

THEOREM 8.5. Problem (3.5) is equivalent to the minimizing problem of the functional

$$F(u^{n+1}) = \frac{1}{2}(D_t\sigma^n, D_t\varepsilon^n) + \frac{1}{2}(\bar{D}_n'(\bar{\varepsilon}^n - \varepsilon^n)/\Delta t, \lambda) - (D_t b^n, D_t u^n),$$

where

$$\lambda = \begin{cases} 0 & \text{for } e \in P^n + E^n \\ \left.\begin{matrix} D_t\varepsilon^n & \text{in } D_+ \\ (\bar{\varepsilon}^n - \varepsilon^n)/\Delta t & \text{in } D_- \end{matrix}\right\} & \text{for } e \in E_p^n \end{cases}$$

under the conditions (A), (B) and (C).

To estimate the error of the approximate solution, we prepare the following theorems ; they correspond to Theorems 8.2 and 8.3 and are proved in the same way.

THEOREM 8.6. Let $(u^n, \sigma^n, \alpha^n)$ $(n = 1 \sim M)$ be the approximate solution obtained by the above procedure.

(i). The following relations hold for any $n \leq M$:

(3.6) $\quad (D_t\varepsilon^n - CD_t\sigma^n, \tau - \sigma^{n+1}) \leq 0 \qquad$ for all $\tau : f(\tau - \alpha^n) \leq z_n$

(3.7) $\quad D_t\alpha^n = \eta S^{-1}(D_t\varepsilon^n - CD_t\sigma^n).$

(ii). Set $\delta^n = z_n - z_0$. Then we have

(3.8) $\quad (\delta^n)^2 \leq \sum_{m=0}^{n-1} f^2([D_t\sigma^m] - D_t\alpha^m)\Delta t^2,$

where $[D_t\sigma^n] = \bar{D}_t\sigma^n$ for the element of E_p^n , and $= D_t\sigma^n$ for that of $E - E_p^n$.

THEOREM 8.7. Let (u, σ, α) be the solution of (3.1)~(3.3) and $(u^n, \sigma^n, \alpha^n)$ the approximate solution obtained above. Then the following a priori estimates hold :

(3.9) $\quad \|\dot{\varepsilon}\|_\infty \, , \|\dot{\sigma}\|_\infty \, , \|\dot{\alpha}\|_\infty \leq C$

(3.10) $$\|D_t\varepsilon^n\|\ ,\ \|D_t\sigma^n\|\ ,\ \|D_t\alpha^n\| \le C,$$

where C is a constant independent of n and Δt, $\|\cdot\|$ is $L^2(\Omega)$ norm of vector function, and $\|\varepsilon\|_\infty = \operatorname{ess\cdot sup}_I \|\varepsilon\|$.

The error estimate of the approximate solution is given by

THEOREM 8.8. Let (u,σ,α) and (u^n,σ^n,α^n) be those in the preceding theorem. Then there is a constant C which is independent of n, Δt and h such that

$$\|\varepsilon(t_n) - \varepsilon^n\|,\ \|\sigma(t_n) - \sigma^n\|,\ \|\alpha(t_n) - \alpha^n\| \le C\Delta t^{\frac{1}{4}} \qquad (n \le M).$$

PROOF. We define $\theta_n = (\sigma^n - \alpha^n)/z_n$ so that $f(\theta_n) \le 1$. As noted in the proof of Theorem 6.9, we have

(3.11) $$\int_I (\dot\varepsilon - C\dot\sigma, \tau - \sigma)dt \le 0 \qquad \text{for all } \tau \in C_+^1(I) : f(\sigma - \alpha) \le z_0$$

(3.12) $$S\dot\alpha = \eta(\dot\varepsilon - C\dot\sigma).$$

Here we can put $\tau = \alpha + \theta_{n+1}z_0$. Since (3.11) holds even when I is replaced by $[t_n,t_{n+1}]$, for example, we can substitute (3.12) into (3.11) and get

$$\begin{aligned}
0 &\ge \int_{t_n}^{t_{n+1}} (S\dot\alpha, \alpha + \theta_{n+1}z_0 - \sigma)dt \\
&= (SD_t\alpha(t_n), \alpha(t_{n+1}) + \theta_{n+1}z_0 - \sigma(t_{n+1}))\Delta t \\
&\quad + \int_{t_n}^{t_{n+1}} (S[\alpha(t_n) - \alpha], \dot\alpha - \dot\sigma)dt \\
&= (SD_t\alpha(t_n), \alpha(t_{n+1}) - \alpha^{n+1} - [\sigma(t_{n+1}) - \sigma^{n+1}])\Delta t \\
&\quad + (SD_t\alpha(t_n), \theta_{n+1}(z_0 - z_{n+1}))\Delta t
\end{aligned}$$

(3.13) $$\quad + \int_{t_n}^{t_{n+1}} (S[\alpha(t_n) - \alpha], \dot\alpha - \dot\sigma)dt.$$

On the other hand, we can set

$$\tau = \alpha^n + \frac{\sigma(t_{n+1}) - \alpha(t_{n+1})}{z_n} z_n$$

in (3.6), and hence we have by (3.7)

$$0 \geq \eta(D_t\varepsilon^n - CD_t\sigma^n, \alpha^n - \alpha(t_{n+1}) + \sigma(t_{n+1}) - \sigma^{n+1})\Delta t$$

$$= (SD_t\alpha^n, \alpha^{n+1} - \alpha(t_{n+1}) + \sigma(t_{n+1}) - \sigma^{n+1})\Delta t$$

$$- (SD_t\alpha^n, D_t\alpha^n)\Delta t^2.$$

Adding this inequality and (3.13), we have

$$0 \geq (SD_t[\alpha(t_n) - \alpha^n], \alpha(t_{n+1}) - \alpha^{n+1} - [\sigma(t_{n+1}) - \sigma^{n+1}])\ \Delta t$$

(3.14) $$+ (SD_t\alpha(t_n), \theta_{n+1}[z_0 - z_{n+1}])\Delta t$$

$$+\int_{t_n}^{t_{n+1}} (S[\alpha(t_n) - \alpha], \dot{\alpha} - \dot{\sigma})dt - (SD_t\alpha^n, D_t\alpha^n)\Delta t^2.$$

The last three terms of the right side are estimated by Theorems 8.6 and 8.7 as follows :

$$|\text{second term}| \leq C\|D_t\alpha(t_n)\|\,\|z_0 - z_{n+1}\|\ \Delta t$$

$$\leq C\|\delta^{n+1}\|\ \Delta t \leq C\ (\sum_m \| D_t\sigma^m\|^2 \cdot \Delta t^2)^{\frac{1}{2}}\Delta t$$

$$\leq C\ \Delta t^{\frac{3}{2}}$$

$$|\text{third term}| \leq C\|\dot{\sigma}\|_\infty^2\ \Delta t^2 \leq C\Delta t^2$$

$$|\text{fourth term}| \leq C\| D_t\alpha^n\|\ \Delta t^2 \leq C\Delta t^2.$$

Summing (3.14) with respect to n, we hence have by (3.7)

$$\| \alpha(t_{n+1}) - \alpha^{n+1}\|^2 + \| \sigma(t_{n+1}) - \sigma^{n+1}\|\ ^2$$

$$\leq C \sum_m \Delta t^{\frac{3}{2}} \leq C\Delta t^{\frac{1}{2}} .$$

This completes the proof of the theorem.

REMARK. Step 4 to determine $(\bar{\sigma}^n, \bar{\varepsilon}^n)$ reduces to seeking a solution of a single algebraic equation of order 6. This is evident since, for a certain θ

$$\bar{\sigma}^n - \sigma^n = \theta S(\bar{\sigma}^n - \alpha^n).$$

Note that this θ is considerablly smaller than unity since σ^n will lie near the boundary of the yield surface (see Fig. 14). The condition that $\bar{\sigma}^n$ lies on the yield surface is expressed by an algebraic equation of θ.

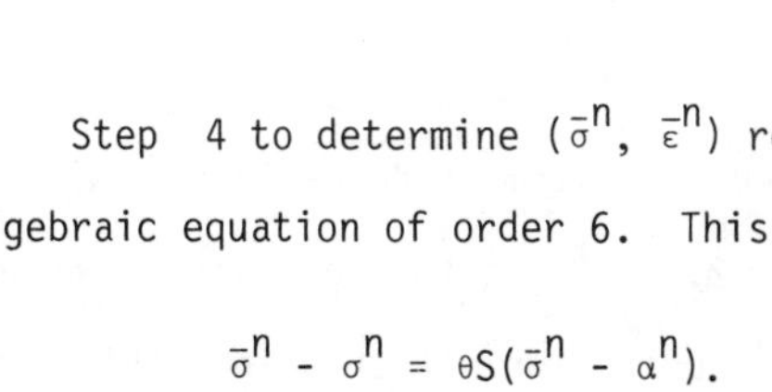

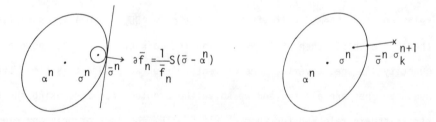

Fig. 14 Determination of $\bar{\sigma}^n$

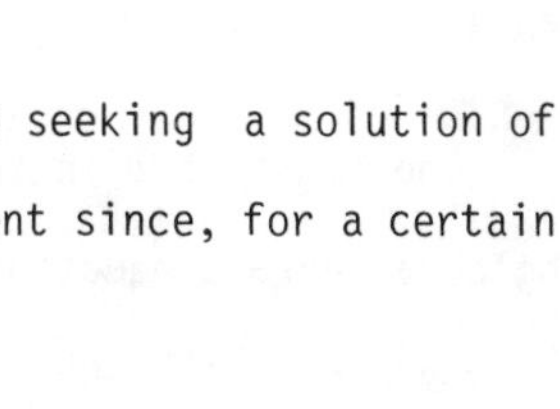

Fig. 15 A modified algorithm

This $(\bar{\sigma}^n, \bar{\varepsilon}^n)$, however, can be replaced by other suitable vector without losing the convergence. One example is to replace step 4 - step 5 by the following step 4' - step 6' (see Fig. 15) :

Step 4'. For $e \in E_0^n$, determine $(\bar{\sigma}^n, \bar{\varepsilon}^n) \in R^3 \times R^3$ satisfying $\bar{\sigma}^n \in S_e^n$ and

$$\bar{\sigma}^n = \sigma^n + \theta(\sigma_k^{n+1} - \sigma^n) \qquad \bar{\varepsilon}^n - \varepsilon^n = C(\bar{\sigma}^n - \sigma^n).$$

Set $E_P^n = E_P^n + E_0^n$, and $\bar{D}'_n = D\partial\bar{f}_n \partial\bar{f}^*_n D/(\eta + \partial\bar{f}^*_n D\partial\bar{f}_n)$, where $\partial\bar{f}_n = \partial f|_{(\bar{\sigma}^n - \alpha^n)}$.

Step 5' = Step 5.

Step 6'. If $\|\bar{\sigma}^n - \sigma^n\| \leq \gamma \|\sigma_{k+1}^{n+1} - \sigma^n\|$ (γ : a constant given a priori), then go to step 6. Otherwise return to step 4 and take the original route, step 4 – step 5.

The constant θ in the modified algorithm is determined as a solution of a single algebraic equation of order 2. As the constant γ we set $\gamma = 10$, for example, although there is no criterion for choosing this since this constant is necessary only for excluding a pathological case in the convergence proof.

REMARK. In solving (3.4) and (3.5) practically, we use the method of trial and error. Consider the case of (3.5). First we assume that all the elements of $P^n + E_p^n$ are plastic and then use the plastic stress-strain relation for them. If $\partial f_n^* D D_t \varepsilon^n \geq 0$ (or $\partial \bar{f}_n^* D \bar{D}_t \varepsilon^n \geq 0$) is satisfied for these elements, then this assumption is correct. If, however, this quantity is negative for some elements, then we assume this time that the negative ones are elastic,and we solve the problem once more using the elastic stress-strain relation for them. If $P^n + E_p^n$ consists of only one element we get the desired solution at this stage. If, however, it includes more than one element, then we might have to continue this trial. Anyway, we can get the desired solution after a finite number of trials since Theorem 8.5 assures the existence of a unique solution. This method is usefull in so far as each element does not change state so often. In fact, this will be assured if the function b has good properties. Also, notice that we can reduce the number of the elements of E_p^n by controlling the step size $\Delta t = t_{n+1} - t_1$. If we control Δt so that this number becomes 1, then our procedure coincides with the well known method of Yamada.

8.4. Isotropic hardening problem

The semidiscrete system based on the isotropic hardening assumption is as follows (see also page 12) :

$$(4.1) \qquad \sum_{j=1}^{2} (\sigma_{ij},\phi_{p,j}) = (b_i,\phi_p) \qquad p \in P$$

$$(4.2) \qquad \begin{cases} \dot{\sigma} = D\dot{\varepsilon} & \text{if } f(\sigma)<H(\bar{\varepsilon}^p) \text{ or } f(\sigma)=H(\bar{\varepsilon}^p) \text{ and } \partial f^*\dot{\sigma} < 0 \\ \dot{\sigma} = (D - D')\dot{\varepsilon} & \text{if } f(\sigma)=H(\bar{\varepsilon}^p) \text{ and } \partial f^*\dot{\sigma} \geq 0, \end{cases}$$

where

$$\bar{\varepsilon}^p = \int_0^t \frac{\sigma^*\dot{\varepsilon}^p}{f(\sigma)}\, dt , \qquad \dot{\varepsilon}^p = \dot{\varepsilon} - C\dot{\sigma} \qquad (C = D^{-1})$$

$$D' = \frac{D\partial f \partial f^* D}{H' + \partial f^* D \partial f}, \qquad H' = H'(H^{-1}(f(\sigma))).$$

We can apply to this problem an algorithm based on the same idea as in the kinematic hardening problem. Below we shall briefly describe the main part of this algorithm. In the present problem, the yield surface of stage n for an element e is of the form

$$S_e^n = \{ \tau \in R^3 \ ; \ f(\tau) = z_n \}.$$

Until the stress point σ reaches the initial yield surface $f(\sigma) = z_n = H(0)$, we employ the elastic stress-strain relation $D_t\sigma^n = DD_t\varepsilon^n$. Assume now that some element satisfies $f(\sigma^n) = z_n = H(0)$.

Step 1. Classify E as $E = P^n + E^n$, where

$$E^n = \{ e \in E \ ; \ f(\sigma^n) < H(0)\}$$

$$P^n = \{ e \in E \ ; \ f(\sigma^n) = H(0)\} .$$

Set $k = 1$, $z_n = H(0)$.

Step 2. Compute $(u_k^{n+1},\sigma_k^{n+1})$ by solving the equation

$$(\sigma_k^{n+1},\phi) = (b^{n+1},\phi) \tag{4.3}$$

under the rule

$$D_t\sigma^n = (D - D_n')D_t\varepsilon^n \qquad \text{in } D_+ = \{\, u^{n+1};\ \partial f_n^*DD_t\varepsilon^n \geq 0 \,\}$$

$$D_t\sigma^n = DD_t\varepsilon^n \qquad \text{in } D_- = \{\, u^{n+1};\ \partial f_n^*DD_t\varepsilon^n < 0 \,\}$$

for P^n, where $D_n' = D\partial f_n \partial f_n^* D/(H_n' + \partial f_n^* D\partial f_n)$ and $H_n' = H'(H^{-1}(z_n))$, and

$$D_t\sigma^n = DD_t\varepsilon^n \qquad \text{to } E^n.$$

Step 3. Classify E^n as $E^n = E_0^n + (\text{new})E^n$, where

$$E_0^n = \{\, e \in E^n \ ;\ f(\sigma_k^{n+1}) \geq z_n \}$$

$$(\text{new})\ E^n = E^n - E_0^n .$$

Set $E_P^n = \Phi$ ..... empty.

Step 4. For $e \in E_0^n$, determine $(\bar\sigma^n,\bar\varepsilon^n) \in R^3 \times R^3$ satisfying $\bar\sigma^n \in S_e^n$ and

$$\| \sigma^n - \bar\sigma^n \|_{R^3} = \operatorname*{Min}_{\sigma \in S_e^n} \| \sigma - \sigma^n \|_{R^3} , \quad \sigma^n - \bar\sigma^n = D(\varepsilon^n - \bar\varepsilon^n).$$

Set $E_P^n = E_P^n + E_0^n$ and $\bar D_n' = D\partial\bar f_n \partial\bar f_n^* D/(H_n' + \partial\bar f_n^* D\partial\bar f_n)$, where $\partial\bar f_n = \partial f|_{\bar\sigma^n}$.

Step 5. Compute $(u_{k+1}^{n+1}, \sigma_{k+1}^{n+1})$ by the same algorithm as in step 5 in the kinematic hardening problem (page 158).

Step 6. Define new E_0^n by $E_0^n = \{\, e \in E^n \ ;\ f(\sigma_{k+1}^{n+1}) \geq z_n \}$.

Step 7. If E_0^n is empty, then go to Step 8. Otherwise define new E^n and k by $E^n = E^n - E_0^n$ and $k = k+1$, and return to Step 4.

Step 8. Define $(u^{n+1},\sigma^{n+1}) = (u_{k+1}^{n+1},\sigma_{k+1}^{n+1})$.

COMMENT. This algorithm reduces finally to solve the equation

(4.4) $(\sigma^{n+1}, \ ) = (b^{n+1}, \phi)$

under the following rule.

(1) $D_t\sigma^n = (D-D_n')D_t\varepsilon^n$ if $f(\sigma^n)=z_n$ and $\partial f_n^* D D_t \varepsilon^n \geq 0$

(2) $\bar{D}_t\sigma^n = (D-\bar{D}_n')\bar{D}_t\varepsilon^n$ if $f(\sigma^n)<z_n$ and $\partial \bar{f}_n^* D \bar{D}_t \varepsilon^n \geq 0$

where $\bar{\sigma}^n - \sigma^n = D(\bar{\varepsilon}^n - \varepsilon^n)$, $f(\bar{\sigma}^n)=z_n$, $\bar{D}_t\sigma^n = (\sigma^{n+1}-\bar{\sigma}^n)/\Delta t$, and

(3) $D_t\sigma^n = DD_t\varepsilon^n$

if $(3)_a$: $f(\sigma^n)$, $f(\sigma^{n+1})<z_n$

or $(3)_b$: $f(\sigma^n)=z_n$ and $\partial f_n^* D D_t \varepsilon^n < 0$

or (3) : $f(\sigma^n)<z_n$ and $\partial \bar{f}_n^* D \bar{D}_t \varepsilon^n < 0$.

Note that the only difference between two algorithms is that $f(\sigma-\alpha)$ is replaced by $f(\sigma)$ in the isotropic case.

Step 9. Determine the yield surface of step n+1.

For cases (1) and (2) : $f(\tau) = z_{n+1} = f(\sigma^{n+1})$.

For case $(3)_a$: $f(\tau) = z_{n+1} = z_n$.

For cases $(3)_b$ and $(3)_c$:

if $f(\sigma^{n+1}) \leq z_n$, then $f(\tau) = z_{n+1} = z_n$,

if $f(\sigma^{n+1}) > z_n$, then $f(\tau) = z_{n+1} = f(\sigma^{n+1})$

We shall introduce ε_n^p and $\bar{\varepsilon}_n^p$ defined as follows ($\varepsilon_0^p = \bar{\varepsilon}_0^p = 0$) :

$D_t\varepsilon_n^p = D_t\varepsilon^n - CD_t\sigma^n$ (or $\bar{D}_t\varepsilon_n^p = \bar{D}_t\varepsilon^n - C\bar{D}_t\sigma^n$ for case (2) above)

$$D_t\bar{\varepsilon}^p_n = \begin{cases} (\sigma^n)^* D_t \varepsilon^p_n / f(\sigma^n) & \text{for cases (1) and (3) above,} \\ (\bar{\sigma}^n)^* \bar{D}_t \varepsilon^p_n / f(\bar{\sigma}^n) & \text{for case (2) above .} \end{cases}$$

THEOREM 8.9. The above procedure, Step 1 → Step 9, is well defined and holds the following a priori estimates of the solution :

$$\|D_t\varepsilon^n\|,\ \|D_t\sigma^n\|\ ,\ \|D_t\bar{\varepsilon}^p_n\| \le C\|D_t b^n\|$$

where C is a constant independent of Δt, n and h.

PROOF. The unique solvability of equations (4.3) and (4.4) is proved in the same manner as in the kinematic hardening problem. Thus the first half of this theorem is evident. To get the a priori estimates we first note that the identity

$$(D_t\sigma^n, D_t\varepsilon^n) = (D_t b^n, D_t u^n)$$

holds for any n. Assume now that the elastic-plastic rule (that is, the relation (2) in the above comment) is applied to get σ^{n+1} for an element e. Then, since

$$(\bar{D}'_n(\bar{\varepsilon}^n - \varepsilon^n)/\Delta t, D_t\varepsilon^n)_e = (\partial\bar{f}^*_n D(\bar{\varepsilon}^n - \varepsilon^n), \partial\bar{f}^*_n D D_t\varepsilon^n)_e / (H'_n + \partial\bar{f}^*_n D \partial\bar{f}_n) \ge 0,$$

we have

$$\begin{aligned}(D_t\sigma^n, D_t\varepsilon^n)_e &= ([D - \bar{D}'_n]D_t\varepsilon^n + \bar{D}'_n(\bar{\varepsilon}^n - \varepsilon^n)/\Delta t, D_t\varepsilon^n)_e \\ &\ge ([D - \bar{D}'_n]D_t\varepsilon^n, D_t\varepsilon^n)_e\end{aligned}$$

from which follows the estimate $\|D_t\varepsilon^n\|_e \le C\|D_t\sigma^n\|_e$. From this other estimates are easily derived.

To show the convergence of the approximate solution, we first list up some of its properties. Let us introduce a parameter $\hat{\varepsilon}^p_n (n = 1,2,\ldots,\ \hat{\varepsilon}^p_0 = 0)$ defined as

$$(4.5)\qquad D_t\hat{\varepsilon}^p_n = \sqrt{H'_n}\, D_t\bar{\varepsilon}^p_n = \begin{cases} \partial f^*_n D_t\sigma^n/\sqrt{H'_n} & \text{for case (1)} \\ \partial\bar{f}^*_n\bar{D}_t\sigma^n/\sqrt{H'_n} & \text{for case (2)} \\ 0 & \text{for case (3)} \end{cases}$$

Also let us introduced $\tilde{\varepsilon}^p_n$ by

$$\tilde{\varepsilon}^p_n = \int_0^{H^{-1}(z_n)} \sqrt{H'(\lambda)}\, d\lambda .$$

Let G, B_G and K_G be those defined in Chapter 5 (see page 100). Notice that if $f(\sigma^n) = z_n$ (or $f(\bar{\sigma}^n) = z_n$), that is, if σ^n (or $\bar{\sigma}^n$) lies on the yield surface S^n_e, then it holds that

$$f(\sigma^n) = G(\tilde{\varepsilon}^p_n) \qquad (\text{ or } f(\bar{\sigma}^n) = G(\tilde{\varepsilon}^p_n)),$$

since $G(\tilde{\varepsilon}^p_n) = H(H^{-1}(z_n)) = z_n = f(\sigma^n)$ (or $= f(\bar{\sigma}^n)$) .

THEOREM 8.10. The solution $(\sigma^n, \varepsilon^n)$ of the present algorithm, and $\hat{\varepsilon}^p_n$ and $\tilde{\varepsilon}^p_n$ defined above satisfy the following relations : $f(\sigma^n) \leq G(\tilde{\varepsilon}^p_n)$ and

$$(4.6)\qquad \sum_{j=1}^{2} (\sigma^{n+1}_{ij}, \phi_{p,j}) = (b^{n+1}_i, \phi_p) \qquad p \in P$$

$$(4.7)\qquad (D_t\varepsilon^p_n, \tau - \sigma^{n+1}) - (D_t\hat{\varepsilon}^p_n, \eta - \tilde{\varepsilon}^p_n) \leq 0 \qquad \text{for all } (\tau,\eta) \in K_G$$

PROOF. It suffices to show the inequality (4.7). According to the COMMENT on page 167 , there are three cases to be checked.

Case (1) : $f(\sigma^n) = z_n = G(\tilde{\varepsilon}^p_n)$, and

$$\begin{pmatrix} D_t\varepsilon_n^p \\ \\ -D_t\hat{\varepsilon}_n^p \end{pmatrix} = \begin{pmatrix} \partial f_n \dfrac{\partial f_n^* D_t\sigma''}{H'} \\ \\ -\sqrt{H_n'}\, D_t\bar{\varepsilon}_n^p \end{pmatrix} = \begin{pmatrix} \partial f_n \\ \\ -\sqrt{H_n'} \end{pmatrix} \frac{\partial f_n^* D_t\sigma^n}{H_n'}$$

Since $G'(\tilde{\varepsilon}_n^p) = \sqrt{H'(H^{-1}(z_n))} = \sqrt{H_n'}$, this shows that the vector $(D_t\varepsilon_n^p, -D_t\hat{\varepsilon}_n^p)$ is parallel to the outward normal to the convex set B_G at the boundary point $(\sigma^n, \tilde{\varepsilon}_n^p)$. Since $(D_t\varepsilon_n^p, \sigma^{n+1} - \sigma^n)$ is nonnegative (4.7) follows.

Case (2) : $f(\bar{\sigma}^n) = z_n$, and

$$\begin{pmatrix} D_t\varepsilon_n^p \\ \\ -D_t\hat{\varepsilon}_n^p \end{pmatrix} = \begin{pmatrix} \partial\bar{f}_n \\ \\ -\sqrt{H_n'} \end{pmatrix} \frac{\partial\bar{f}_n^* \bar{D}_t\sigma^n}{H_n'}$$

Hence the vector $(D_t\varepsilon_n^p, -D_t\hat{\varepsilon}_n^p)$ is again parallel to the outward normal to B_G at the point $(\bar{\sigma}^n, \tilde{\varepsilon}_n^p)$.

Finally this vector is zero in case(3). Hence inequality (4.7) holds well.

The difference between $\hat{\varepsilon}^p$ and $\tilde{\varepsilon}^p$ is caused by the time discretization. It is estimated by

THEOREM 8.11.

(4.8) $$0 \le D_t\tilde{\varepsilon}_n^p - D_t\hat{\varepsilon}_n^p \le C\|D_t\sigma^n\|^2_{R^3}\,\Delta t.$$

PROOF. To prove the first inequality, it suffices to show the relation

$$\int_{H^{-1}(z_n)}^{H^{-1}(z_{n+1})} \sqrt{H'(\lambda)}\,d\lambda \ge \frac{1}{\sqrt{H'(H^{-1}(z_n))}}\,\partial f_n^*\, D_t\sigma^n\Delta t$$

under the assumptions that $\partial f_n^* D_t\sigma^n \ge 0$, $z_{n+1} = f(\sigma^{n+1}) \ge f(\sigma^n) = z_n$. Let β be defined by

$$\beta = H(\lambda) \qquad\qquad H^{-1}(z_n) \le \lambda \le H^{-1}(z_{n+1}).$$

Then we have the desired inequality

$$\int_{H^{-1}(z_n)}^{H^{-1}(z_{n+1})} \sqrt{H'(\lambda)}\, d\lambda = \int_{z_n}^{z_{n+1}} \frac{1}{\sqrt{H'(H^{-1}(\beta))}}\, d\beta \ \ge \ \frac{\partial f_n^* D_t \sigma^n \Delta t}{\sqrt{H'(H^{-1}(z_n))}}$$

provided the following inequality holds :

$$(4.9) \qquad z_{n+1} - z_n \ge \partial f_n^* D_t \sigma^n \Delta t.$$

This inequality is proved as follows : Since $f_n \partial f_n = S\sigma^n$, we have

$$\{ \partial f_n^*(\sigma^{n+1} - \sigma^n)\}^2 = \{ \frac{(S\,\sigma^n, \sigma^{n+1} - \sigma^n)_{R^3}}{f_n} \}^2$$

$$= \{ \frac{(\sqrt{S}\sigma^n, \sqrt{S}(\sigma^{n+1} - \sigma^n))_{R^3}}{f_n} \}^2$$

$$\le \|\sqrt{S}(\sigma^{n+1} - \sigma^n)\|_{R^3}^2 = f^2(\sigma^{n+1} - \sigma^n).$$

Hence $0 \le \partial f_n^* D_t \sigma^n \le f(D_t \sigma^n)$, and we have

$$z_{n+1} = f(\sigma^{n+1}) = z_n \sqrt{1 + \frac{2\partial f_n^* D_t \sigma^n \Delta t}{z_n} + \frac{f^2(D_t\sigma^n)\Delta t^2}{z_n^2}}$$

$$\ge z_n + \partial f_n^* D_t \sigma^n \Delta t.$$

To prove the second inequality, we check the three cases in the previous COMMENT.

Case (1) : We have

$$(D_t \tilde{\varepsilon}_n^p - D_t \hat{\varepsilon}_n^p)\Delta t = \int_{H^{-1}(z_n)}^{H^{-1}(z_{n+1})} \sqrt{H'(\lambda)}\, d\lambda - \frac{1}{\sqrt{H_n'}}\, \partial f_n^* D_t \sigma^n \Delta t$$

$$= \int_{z_n}^{z_n'} \left[\frac{1}{\sqrt{H'(H^{-1}(\beta))}} - \frac{1}{\sqrt{H'(H^{-1}(z_n))}} \right] d\beta + O(|\delta^n|)$$

where $z_n' = z_n + \partial f_n^* D_t \sigma^n \Delta t$ and $|\delta^n| \le C\|D_t \sigma^n\|_{R^3}^2 \Delta t^2$. Since $0 \le \partial f_n^* D_t \sigma^n \Delta t$

$\leq f(D_t\sigma^n)\Delta t$ and, for $\beta \in [z_n, z_{n+1}]$,

$$1/\sqrt{H'(H^{-1}(\beta))} - 1/\sqrt{H'(H^{-1}(z_n))}$$

$$\leq 1/\sqrt{H'(H^{-1}(z_{n+1}))} - 1/\sqrt{H'(H^{-1}(z_n))}$$

$$\leq C\,(f(\sigma^{n+1}) - f(\sigma^n)) \leq C\|D_t\sigma^n\|_{R^3}\,\Delta t,$$

we have the desired estimate .

Case (2) : Similarly we have

$$(D_t\tilde{\varepsilon}^p_n - D_t\hat{\varepsilon}^p_n)\Delta t \leq C\|\bar{D}_t\sigma^n\|^2_{R^3}\Delta t^2 \leq C\|D_t\sigma^n\|^2_{R^3}\Delta t^2.$$

Case (3) : In this case , $D_t\hat{\varepsilon}^p_n = 0$, and $D_t\tilde{\varepsilon}^p_n = 0$ or else

$$D_t\tilde{\varepsilon}^p_n\Delta t = \int_{H^{-1}(z_n)}^{H^{-1}(z_{n+1})} \sqrt{H'(\lambda)}\, d\lambda = \int_{z_n}^{z_{n+1}} \frac{1}{\sqrt{H'(H^{-1}(\beta))}}\, d\beta$$

$$\leq C(f(\sigma^{n+1}) - f([\sigma^n]))$$

$$\leq C \sum_{|\alpha|=2} h^\alpha\left(\frac{\partial}{\partial\sigma}\right) f([\sigma^n] + \theta h) \leq C\|D_t\sigma^n\|^2_{R^3}\Delta t^2 \quad (0<\theta<1),$$

where $h = \sigma^{n+1} - [\sigma^n]$ and $[\sigma^n] = \sigma^n$ or $= \bar{\sigma}^n$.

Hence the second estimate holds in any case.

For estimating the error of the approximate solution we prepare

THEOREM 8.12. Let $x_n (n = 1,2,\dots)$ be a sequence of nonnegative numbers satisfying

$$x_n^2 \leq C_1 + C_2 x_n + C_3 \sum_{k=0}^{n-1} x_k \Delta t \qquad (x_0 = 0), \tag{4.10}$$

where C_i are positive constants. Then it holds that

$$x_n \leq \frac{C_2}{2} + \sqrt{d} + \frac{C_3}{2} n\Delta t \tag{4.11}$$

for any n ($n\Delta t \leq I$), where

$$d = C_1 + \frac{C_2^2}{4} + \frac{C_2C_3}{2}I .$$

PROOF. Set $y_k = |x_k - C_2/2|$. Then y_k satisfies

$$y_n^2 \leq d + C_3 \sum_{k=0}^{n-1} y_k \Delta t.$$

Let { z_k } be the sequence satisfying $z_0 = y_0$, $z_n^2 = d + C_3 \sum_{k=0}^{n-1} z_k \Delta t$ ($n \geq 1$). Then we have $y_n \leq z_n$ for any $n \geq 1$. To see this, we use an induction on n. Clearly this is correct for n = 1. Assume $y_k \leq z_k$ for any $k \leq n$ ($n \geq 1$). We then have

$$y_{n+1}^2 \leq d + C_3 \sum_{k=0}^{n} y_k \Delta t \leq d + C_3 \sum_{k=0}^{n} z_k \Delta t = z_{n+1}^2,$$

which completes the induction. We next consider the function z(t) defined by

$$z(t)^2 = d + C_3 \int_0^t z(s)\, ds \qquad (t \geq 0). \tag{4.12}$$

Then we have $z_n \leq z(n\Delta t)$ ($n \geq 0$). To see this, we again use an induction. Clearly $z_0 = y_0 = C_2/2 \leq \sqrt{d} = z(0)$. Assume that $z_k \leq z(k\Delta t)$ holds for all $k \leq n$. Since z(t) is increasing, we have

$$\begin{aligned} z_{n+1}^2 &= d + C_3 \sum_{k=0}^{n} z_k \Delta t \leq d + C_3 \sum_{k=0}^{n} z(k\Delta t)\Delta t \\ &\leq d + C_3 \sum_{k=0}^{n} \int_{k\Delta t}^{(k+1)\Delta t} z(s)\, ds \\ &= d + C_3 \int_0^{(n+1)\Delta t} z(s)\, ds = z^2((n+1)\Delta t). \end{aligned}$$

Hence z_{n+1} is bounded by $z((n+1)\Delta t)$. Now (4.12) can be solved easily. In fact, we have $z(t) = \sqrt{d} + C_3/2\cdot t$. This completes the proof of the theorem.

THEOREM 8.13. Let $(u, \sigma, \bar{\varepsilon}^p)$ be the solution of the semidiscrete system (4.1)$\sim$(4.2), and set

$$E^2(n) = (C[\sigma(n) - \sigma^n],\sigma(n) - \sigma^n) + (\hat{\varepsilon}^p(n) - \hat{\varepsilon}^p_n,\hat{\varepsilon}^p(n) - \hat{\varepsilon}^p_n),$$

where $\sigma(n) = \sigma(n\Delta t)$, and

$$\hat{\varepsilon}^p(n) \equiv \hat{\varepsilon}^p(n\Delta t) = \int_0^{\bar{\varepsilon}^p(n\Delta t)} \sqrt{H'(\lambda)}\, d\lambda .$$

Then we have the error estimate

$$E(n) \le C\sqrt{\frac{\Delta t}{h}} \qquad (n\Delta t \le I), \tag{4.13}$$

where C is a constant independent of n, Δt and h.

PROOF. The idea of the proof is the same as in the kinematic hardening case. We set $\tau = \sigma^{n+1}$ and $\zeta = \tilde{\varepsilon}^p_{n+1}$ in the inequality (4.7) of Chapter 6 (see page 120). Then we integrate in $I_n = (n\Delta t,(n+1)\Delta t)$ to get

$$\begin{aligned} 0 &\ge \int_{I_n} (\dot{\varepsilon}^p,\sigma^{n+1} - \sigma)dt - \int_{I_n} (\dot{\hat{\varepsilon}}^p,\tilde{\varepsilon}^p_{n+1} - \hat{\varepsilon}^p)dt \\ &= (D_t\varepsilon^p(n),\sigma^{n+1} - \sigma(n+1))\Delta t + \int_{I_n} (\varepsilon^p - \varepsilon^p(n),\dot{\sigma})dt \\ &\quad - (D_t\hat{\varepsilon}^p(n),\tilde{\varepsilon}^p_{n+1} - \hat{\varepsilon}^p(n+1))\Delta t - \int_{I_n} (\hat{\varepsilon}^p - \hat{\varepsilon}^p(n),\dot{\hat{\varepsilon}}^p)dt. \end{aligned}$$

On the other hand, we have by Theorem 8.10

$$\begin{aligned} 0 &\ge (D_t\varepsilon^p_n,\sigma(n+1) - \sigma^{n+1})\Delta t - (D_t\hat{\varepsilon}^p_n,\hat{\varepsilon}^p(n+1) - \tilde{\varepsilon}^p_n)\Delta t \\ &= (D_t\varepsilon^p_n,\sigma(n+1) - \sigma^{n+1})\Delta t - (D_t\hat{\varepsilon}^p_n,\hat{\varepsilon}^p(n+1) - \tilde{\varepsilon}^p_{n+1})\Delta t \\ &\quad + (D_t\hat{\varepsilon}^p_n,\tilde{\varepsilon}^p_n - \tilde{\varepsilon}^p_{n+1})\Delta t. \end{aligned}$$

Adding both inequalities, we have

$$0 \geq (D_t[\varepsilon^p(n) - \varepsilon^p_n], \sigma^{n+1} - \sigma(n+1))\Delta t$$

$$+ (D_t[\hat{\varepsilon}^p(n) - \hat{\varepsilon}^p_n], \hat{\varepsilon}^p(n+1) - \tilde{\varepsilon}^p_{n+1})\Delta t$$

$$+ \int_{I_n} (\varepsilon^p - \varepsilon^p(n), \dot{\sigma})dt - \int_{I_n} (\hat{\varepsilon}^p - \hat{\varepsilon}^p(n), \dot{\hat{\varepsilon}}^p)dt$$

$$- (D_t\hat{\varepsilon}^p_n, D_t\tilde{\varepsilon}^p_n)\Delta t^2.$$

The sum of the last 3 terms of the right side of this inequality is bounded by $C\Delta t^2$ as is seen from the a priori estimates of the solution. Hence, summing on n, we have

$$(C[\sigma(n+1) - \sigma^{n+1}], \sigma(n+1) - \sigma^{n+1}) + (\hat{\varepsilon}^p(n+1) - \hat{\varepsilon}^p_{n+1}, \hat{\varepsilon}^p(n+1) - \hat{\varepsilon}^p_{n+1})$$

$$+ 2\sum_{k=0}^{n} (D_t[\hat{\varepsilon}^p(k) - \hat{\varepsilon}^p_k], \hat{\varepsilon}^p_{k+1} - \tilde{\varepsilon}^p_{k+1})\Delta t \leq C\Delta t. \tag{4.14}$$

Now, since we have by Theorem 8.11 and the boundedness of $\|D_t\varepsilon^p_n\|_\Omega$

$$(D_t\tilde{\varepsilon}^p_n - D_t\hat{\varepsilon}^p_n)^2 \leq (D_t\tilde{\varepsilon}^p_n)^2 - (D_t\hat{\varepsilon}^p_n)^2$$

$$\leq 2(D_t\tilde{\varepsilon}^p_n)(D_t\tilde{\varepsilon}^p_n - D_t\hat{\varepsilon}^p_n) \leq \frac{C}{h}\|D_t\tilde{\varepsilon}^p_n\|_\Omega \|D_t\sigma^n\|^2_{R^3}\Delta t \leq \frac{C}{h}\|D_t\sigma^n\|^2_{R^3}\Delta t,$$

the last term on the right side of (4.14) is bounded by

$$2|(\hat{\varepsilon}^p(n+1) - \hat{\varepsilon}^p_{n+1}, \hat{\varepsilon}^p_{n+1} - \tilde{\varepsilon}^p_{n+1})| + 2\sum_{k=1}^{n} |(\hat{\varepsilon}^p(k) - \hat{\varepsilon}^p_k, D_t\hat{\varepsilon}^p_k - D_t\tilde{\varepsilon}^p_k)|\,\Delta t$$

$$\leq C\sqrt{\frac{\Delta t}{h}}\,\|\hat{\varepsilon}^p(n+1) - \hat{\varepsilon}^p_{n+1}\| + C\sqrt{\frac{\Delta t}{h}}\sum_{k=1}^{n}\|\hat{\varepsilon}^p(k) - \hat{\varepsilon}^p_k\|\Delta t.$$

Hence, we finally have the inequality

$$E^2(n) \leq C\Delta t + C\sqrt{\frac{\Delta t}{h}}\,E(n) + C\sqrt{\frac{\Delta t}{h}}\sum_{k=0}^{n} E(k)\Delta t,$$

from which (4.13) follows by Theorem 8.12.

CHAPTER 9

ELASTIC-PLASTIC DEFORMATION OF CONTINUOUS BODIES

9.1. Some theorems in the theory of integration and functional analysis

In this section we list up some theorems which are used in this chapter to prove the existence of fully continuous solutions. We omit the proofs since they require considerable preparations. If the reader wants the proof, see [1],[12], for example.

Let H be a Hilber space with inner product (,) and norm $\|\cdot\|$. We say that a sequence $u_n \in H$ $(n = 1,2,\dots)$ converges to $u \in H$ weakly if

$$(u_n,\phi) \to (u,\phi) \qquad \text{for any } \phi \in H.$$

In this case $\underline{\lim}\, \|u_n\| \geq \|u\|$ holds. If u_n converges to u weakly and $\|u_n\|$ converges to $\|u\|$, then u converges to u strongly ; that is $\|u - u_n\| \to 0$.

The first fundamental theorem used for existence proof is

THEOREM 9.1. Let $u_n \in H$ be a uniformly bounded sequence. Then $\{u_n\}$ has a weakly convergent subsequence.

To prove the convergence of the initial and boundary values, we use

THEOREM 9.2. Let H_1 and H_2 be Hilbert spaces, and $\gamma : H_1 \to H_2$ be a continuous linear operator. If $\{u_n\}$ converges to u weakly in H_1, then $\{\gamma u_n\}$ converges to γu weakly in H_2.

We introduce Banach spaces of mappings with values in a Banach space. Let I be a bounded interval (0,T), and let E be a Banach space with norm $\|\cdot\|$.

DEFINITION. A mapping $u : I \to E$ is said to be measurable, if there exists a sequence of step mappings $\{u_n\}$ with respect to the Lebesgue measure on I such that $\{u_n\}$ converges to u almost everywhere on I.

Let $L^1(I ; E)$ be the set of all the measurable mappings $u : I \to E$ such that there exists a sequence of step mappings $u_n : I \to E$ which converges to u almost everywhere on I and which is L^1-Cauchy in the sense that for given $\varepsilon > 0$ there exists n_0 such that $\int \|u_n - u_m\| < \varepsilon$ for all $n,m \geq n_0$. We define

$$\int_I u \, dt = \lim_n \int_I u_n \, dt$$

$$\int_I \|u\| \, dt = \lim_n \int_I \|u_n\| \, dt,$$

where $\{u_n\}$ is a L^1- Cauchy sequence of step mappings which converges to u almost everywhere on I.

DEFINITION. By $L^p(I;E)$ $(1 \leq p < \infty)$ we denote the equivalence classes of mappings $u : I \to E$ such that $\|u\|^p \in L^1(I; R)$ and that differ only on a set of measure zero. The norm of this Banach space is denoted by $\|u\|_p$. $L^\infty(I; E)$ denotes the equivalence classes of measurable mappings $u : I \to E$ such that

$$\|u\|_\infty = \text{essential} \sup_I \|u\| < \infty .$$

The next theorem also is used to prove the existence of the solution.

THEOREM 9.3. Let H be a separable Hilbert space. If $u_n \in L^\infty(I; H)$ satisfies $\|u_n\|_\infty \leq C$ (for all $n \geq 1$) for a certain constant C, there exists a subsequence

$\{u_{n'}\}$ and $u \in L^{\infty}(I; H)$ such that

$$\int_I (u_{n'}, \phi)_H dt \rightarrow \int_I (u, \phi)_H dt \qquad \text{for any } \phi \in L^1(I; H)$$

as $n' \to \infty$.

The above theorem is proved by combining two facts : (1) If X is a separable Banach space and $\{f_n\}$ is a uniformly bounded sequence of bounded linear functionals on X, then $\{f_n\}$ has a weakly convergent subsequence. (2) $L^{\infty}(I; H) \cong$ the dual of $L^1(I; H)$. We shall write the convergence of this theorem as

$$u_{n'} \to u \qquad \text{weakly* in } L^{\infty}(I; H).$$

In the practical applications of these theorems, the spaces E and H are Sobolev spaces of functions. So let us pick up some theorems related to these spaces for later use. Let Q be a bounded domain in R^n. We say that the function $v_i \in L^1(Q)$ is a weak derivative of $u \in L^1(Q)$ if the following equality holds for any $\phi \in D(Q)$ - the set of all C^{∞} functions with support in Q.

$$\int_Q v_i \phi dx = - \int_Q u \frac{\partial \phi}{\partial x_i} dx.$$

In what follows we use the same notation $\partial u/\partial x_i$ to denote the weak and the ordinal derivatives. Also, $\dot{u}$ denotes the derivative with respect to the time in that sense. By $W_2^1(Q)$ we denote the set of all $L^2(Q)$ functions with weak derivatives $\partial u/\partial x_i \in L^2(Q)$ ($i = 1,2,..,n$). $W_2^1(Q)$ is then a Hilbert space by the following inner product :

$$(u,v)_{W^1(Q)} = (u,v)_{L^2(Q)} + \sum_{i=1}^{n} (\frac{\partial u}{\partial x_i}, \frac{\partial v}{\partial x_i})_{L^2(Q)} .$$

The following theorem explains the relation between the weak and ordinal derivatives. We shall use this fact at a key point to reproduce the stress-strain relation from a weak formulation.

THEOREM 9.4. Let $v_1 \in L^1(Q)$ be the weak derivative of $u \in L^1(Q)$. By modifying the value of u on a set of measure zero in Q, if that is necessary, we may regard u(x) as an absolutely continuous function with respect to x_1 for almost all $(x_2, x_3, \ldots, x_n)$, and v_1 is equal to the ordinal derivative $\partial u/\partial x_1$ almost everywhere on Q.

The function $u \in W_2^1(I \times \Omega)$ has its boundary value (trace). To avoid pathological cases in treating the boundary, we reqire the piecewise smoothness of the boundary of Ω.

DEFINITION. We say that the boundary $\partial\Omega$ of a bounded domain Ω in (x_1, x_2) plane has the piecewise C^m-property $(m \geq 1)$, if the following conditions are satisfied :

(1). $\partial\Omega$ consists of a finite number of non-overlapping C^m curves $\{\Gamma_i\}$. Each Γ_i is represented in the form

(A) $\quad x_2 = \phi_i(x_1) \qquad (x_1 \in I_i) \qquad \sup_{I_i} \sum_{\alpha \leq m} \left| \frac{d^\alpha}{dx_1^\alpha} \phi_i \right| < \infty$

or

(B) $\quad x_1 = \phi_j(x_2) \qquad (x_2 \in I_j) \qquad \sup_{I_j} \sum_{\alpha \leq m} \left| \frac{d^\alpha}{dx_2^\alpha} \phi_j \right| < \infty .$

The sign of $x_i - \phi_k(x_j)$ distinguishes the inside and outside of Ω.

(2). For any boundary point p there are triangles with vertex p, one of which is in Ω and another outside Ω . A portion of $\partial\Omega$ is said to be of (A) type (or (B) type) if it is represented as (A) (or (B)).

If the boundary satisfies the above assumptions, then the proof of Korn's inequality becomes very easy (see Appendix A). This is another reason we put forth the above assumption.

The function $u \in W_2^1(Q)$ has its boundary value.

THEOREM 9.5. Let the boundary $\partial\Omega$ of the domain Ω be of piecewise C^1 property, and set $Q = I \times \Omega$.

(1). $C^1(\bar{Q})$ is dense in $W_2^1(Q)$.

(2). The trace of $u \in W_2^1(Q)$ on ∂Q can be defined as a function of $L^2(\partial Q)$, and there is a constant C such that for all $u \in W_2^1(Q)$ it holds that

$$\|u\|_{L^2(\partial Q)} \leq C\|u\|_{W_2^1(Q)} .$$

The following theorems are frequently used.

THEOREM 9.6. (1). u belongs to $L^\infty(I;W_2^1(\Omega))$ if and only if u and $\partial u/\partial x_i$ belong to $L^\infty(I;L^2(\Omega))$.

(2). Let u be a measurable function on $Q = I \times \Omega$. If, for almost all $t \in I$, $u(t,x) = 0$ almost everywhere on Ω, then $u = 0$ almost everywhere on Q, and vice versa.

(3). Let $u \in L^1(Q)$, and let $S_h(t)$ be the closed cube of side length h which is centered at $t \in Q$. Then it holds that

$$\lim_{h \to 0} \frac{1}{\text{measure}(S_h(t))} \int_{S_h(t)} u(x)\, dx = u(t) \qquad \text{a.e. } Q.$$

THEOREM 9.7. Let u and $\dot{u}$ be in $L^2(I\times\Omega)$. Then $u(0,x)$ is well defined as a function of $L^2(\Omega)$, and it holds that for any u

$$\| u(0,x)\|_{L^2(\Omega)} \leq C(\| u \|_{L^2(I\times\Omega)} + \|\dot{u}\|_{L^2(I\times\Omega)}).$$

9.2. Elastic-plastic vibration of a rod

In Chapter 5 we introduced an initial value problem for a system of equations which expresses the vibration of a rod submitted to longitudinal impact. In this section we prove that this problem has a unique solution and that it is obtained as a limit of the solutions of the multiple system considered in the same chapter.

For $\sigma, \alpha \in L^2(I\times\Omega)$ ($I=(0,T), \Omega=(0,1)$) satisfying

$$|\sigma - \alpha| \leq z_0 \qquad \text{a.e. } I\times\Omega ,$$

we define two sets E and P by

$$E = E_{\sigma,\alpha} = \{ (t,x) \in I\times\Omega \; ; \; |\sigma - \alpha| < z_0 \}$$

$$P = P_{\sigma,\alpha} = \{ (t,x) \in I\times\Omega \; ; \; |\sigma - \alpha| = z_0 \},$$

where both σ and α are assumed to be bounded at the above (t,x). Clearly both E and P are measurable and measure($I\times\Omega - E - P$) = 0.

DEFINITION. We say that (u,σ,α) is a solution of a rod problem if

$$u, \dot{u}, \sigma \in L^\infty(I; W_2^1(\Omega)), \qquad \ddot{u}, \dot{\sigma}, \alpha, \dot{\alpha} \in L^\infty(I; L^2(\Omega))$$

$$u(0,x) = \sigma(0,x) = \alpha(0,x) = 0, \quad \dot{u}(0,x) = a(x) \qquad \text{a.e. } \Omega$$

$$u(t,0) = \sigma(t,1) = 0 \qquad \text{a.e. } I,$$

and if the following equations are satisfied :

(2.1) $\rho\ddot{u} - \sigma_x = b$ a.e. $I \times \Omega$

(2.2) $\begin{cases} \dot{\sigma} = k\dot{u}_x & \text{a.e. } E \\ \dot{\sigma} = (1-\xi)k\dot{u}_x \text{ and } \frac{\sigma-\alpha}{|\sigma-\alpha|}\dot{\sigma} \geq 0 & \text{a.e. } P \end{cases}$

(2.3) $\dot{\alpha} = \frac{k(1-\xi)}{\xi}(\dot{u}_x - \frac{1}{k}\dot{\sigma})$ a.e. $I \times \Omega$

(2.4) $|\sigma - \alpha| \leq z_0$ a.e. $I\times\Omega$.

The functions b and a are assumed to satisfy the conditions in Chapter 1.

Now the purpose of this section is to prove

THEOREM 9.8. The solution of the rod problem exists uniquely. It is given as a limit of the finite element solutions.

PROOF. The uniqueness is proved easily. Let (u, σ, α) be a solution of this problem. Since we have

$$\dot{\sigma} - k\dot{u}_x = \begin{cases} 0 & \text{a.e. } E \\ -\xi k\dot{u}_x \quad \text{and} \quad \frac{\sigma-\alpha}{|\sigma-\alpha|}\dot{u}_x \geq 0 & \text{a.e. } P, \end{cases}$$

it holds by (2) of Theorem 9.6 that a.e. on I

(2.5) $$(\dot{\sigma} - k\dot{u}_x, \tau - \sigma)_{L^2(\Omega)} \geq 0$$

for fixed $\tau \in K_\alpha$, where

$$K_\alpha = \{ \tau \in L^\infty(I; L^2(\Omega)) \ ; \ |\tau - \alpha| \leq z_0 \ \text{ a.e. } I\times\Omega\} .$$

Now assume (u', σ', α') is another solution. Take a function θ satisfying

$$\theta \in L^\infty(I; L^2(\Omega)), \quad |\theta| \le 1 \qquad \text{a.e. } I\times\Omega.$$

Then the functions

$$\tau = \alpha + z_0\theta, \qquad \tau' = \alpha' + z_0\theta$$

belong to K_α and $K_{\alpha'}$, respectively, so that (2.3) and (2.5) imply a.e. on I

(2.6) $$(\dot\alpha, \alpha - \sigma + z_0\theta)_{L^2(\Omega)} \le 0,$$

(2.7) $$(\dot\alpha', \alpha' - \sigma' + z_0\theta)_{L^2(\Omega)} \le 0.$$

Here we can set $\theta = (\sigma' - \alpha')/z_0$ in (2.6) and $\theta = (\sigma - \alpha)/z_0$ in (2.7) to get

(2.8) $$(\dot\alpha - \dot\alpha', \alpha - \alpha' - (\sigma - \sigma'))_{L^2(\Omega)} \le 0.$$

By (2.3) we have a.e. on I

$$(\dot\alpha - \dot\alpha', \sigma - \sigma')_{L^2(\Omega)} = \frac{k(1-\xi)}{\xi}\,(\dot u_x - \dot u'_x - \frac{1}{k}(\dot\sigma - \dot\sigma'), \sigma - \sigma')_{L^2(\Omega)},$$

and by (2.1) a.e. on I

$$(\dot u_x - \dot u'_x, \sigma - \sigma')_{L^2(\Omega)} = -\rho(\dot u - \dot u', \ddot u - \ddot u')_{L^2(\Omega)}.$$

Hence (2.8) implies a.e. on I

$$\frac{d}{dt}\left(\frac{1}{2}\|\alpha - \alpha'\|^2_{L^2(\Omega)} + \frac{k(1-\xi)}{2\xi}\rho\|\dot u - \dot u'\|^2_{L^2(\Omega)} + \frac{1-\xi}{2\xi}\|\sigma - \sigma'\|^2_{L^2(\Omega)}\right) \le 0,$$

from which the uniqueness follows.

To prove the existence of the solution, we use the multiple system (1.10)∼(1.11) of Chapter 5. We regard this system as an approximate equation to the original one and use the compactness argument to get a solution of the original problem.

By the analysis in Chapter 3, we already know that this approximate system has a unique solution $(\hat{u},\bar{\bar{\sigma}},\bar{\bar{\alpha}})$ and that the a priori estimates

$$\|\dot{\hat{u}}\|_{W^1(\Omega)}, \ \|\ddot{\bar{\bar{u}}}\|_{L^2(\Omega)}, \ \|\dot{\bar{\bar{\sigma}}}\|_{L^2(\Omega)}, \ \|\dot{\bar{\bar{\alpha}}}\|_{L^2(\Omega)} \leq C$$

hold uniformly on h and t for a certain constant C. Therefore, by Theorem 9.3, there are (u,σ,α) such that for a suitable subsequence

$$\hat{u},\ \dot{\hat{u}},\ \hat{u}_x,\ \dot{\hat{u}}_x,\ \ddot{\bar{\bar{u}}} \ \to \ u,\ \dot{u},\ u_x,\ \dot{u}_x,\ \ddot{u} \qquad \text{weakly* in } L^\infty(I;L^2(\Omega))$$

$$(2.9)\qquad \begin{cases} \bar{\bar{\sigma}},\ \dot{\bar{\bar{\sigma}}} \ \to \ \sigma,\ \dot{\sigma} \\ \bar{\bar{\alpha}},\ \dot{\bar{\bar{\alpha}}} \ \to \ \alpha,\ \dot{\alpha} \end{cases} \qquad \text{weakly* in } L^\infty(I;\ L^2(\Omega))$$

as $h \to 0$. Hence by Theorem 9.6

$$\hat{u},\ \dot{\hat{u}} \ \to \ u,\ \dot{u} \qquad \text{weakly* in } L^\infty(I;\ W_2^1(\Omega)) .$$

It is also clear that

$$\hat{u},\ \dot{\hat{u}} \ \to \ u,\ \dot{u} \qquad \text{weakly in } W_2^1(I\times\Omega).$$

Therefore, by Theorems 9.2 and 9.5, the initial and boundary values of $\hat{u}$ converge to those of u weakly in L^2. Since $\hat{u}(0,x) = \hat{u}(t,0) = 0$, and $\dot{\hat{u}}(0,x)$ converges to a(x) uniformly in Ω, u satisfies the initial and boundary values in the sense of the trace. $\sigma(0,x) = \alpha(0,x) = 0$ follows from $\bar{\bar{\sigma}}(0,x) = \bar{\bar{\alpha}}(0,x) = 0$ and Theorem 9.7.

We introduce a family of functions,

$$K_{\bar{\bar{\alpha}}} = \{\bar{\bar{\tau}} = \sum_i \tau_i(t)\bar{\bar{\phi}}_i(x)\ ;\ \ \tau_i \in C_+^1(I),\ \ \max_I |\bar{\bar{\tau}} - \bar{\bar{\alpha}}| \leq z_0\},$$

where $\bar{\bar{\phi}}_i$ is the characteristic function of the element $[(i-1)h,ih]$. As shown in the proof of Theorem 5.8, the approximate system (1.10)~(1.11) of Chapter 5 satisfies the following relations :

$$(2.10)\qquad \rho(\ddot{\bar{u}},\bar{\phi}_i) + (\bar{\bar{\sigma}},\hat{\phi}_{i,x}) = (b,\hat{\phi}_i) \qquad i = 1 \sim N,$$

$$(2.11)\qquad (\dot{\bar{\bar{\sigma}}} - k\dot{\hat{u}}_x, \bar{\bar{\tau}} - \bar{\bar{\sigma}}) \geq 0 \qquad \text{for all } \bar{\bar{\tau}} \in K_{\bar{\bar{\alpha}}},$$

$$(2.12)\qquad \dot{\bar{\bar{\alpha}}} = \frac{\xi-1}{\xi}(\dot{\bar{\bar{\sigma}}} - k\dot{\hat{u}}_x),$$

where (,) denotes $L^2(\Omega)$ inner product.

PROOF of (2.1). Take $\phi \in C^\infty(\overline{I\times\Omega})$ satisfying $\phi(t,0) = 0$, and consider its approximations

$$\bar{\phi} = \sum_{i=1}^{N} \phi(t,ih)\bar{\phi}_i$$

$$\hat{\phi} = \sum_{i=1}^{N} \phi(t,ih)\hat{\phi}_i .$$

We then have $\|\phi - \bar{\phi}\|_{L^2(I\times\Omega)} \to 0$ and $\|\phi - \hat{\phi}\|_{W^1(I\times\Omega)} \to 0$ as h tends to 0. Substituting these functions into (2.10) we have

$$\int_I [\,\rho(\ddot{u},\phi) + (\sigma,\phi_x) - (b,\phi)]dt$$

$$= \int_I [\,\rho(\ddot{u} - \ddot{\bar{u}},\phi) + (\sigma - \bar{\bar{\sigma}},\phi_x)]\,dt + \varepsilon_h \to 0$$

as $h \to 0$. Since $\ddot{u}$, $b \in L^\infty(I;L^2(\Omega))$, this implies $\sigma_x \in L^\infty(I;L^2(\Omega))$, and hence (2.1) holds. It is also clear that $\sigma(t,1) = 0$.

The equality (2.3) follows obviously from (2.12). To prove (2.4), we first note that $|\bar{\bar{\sigma}} - \bar{\bar{\alpha}}| \leq z_0$ holds in $I\times\Omega$. Let χ be the characteristic function of an measurable set in $I\times\Omega$. Then, since we can suppose that $\bar{\bar{\sigma}} - \bar{\bar{\alpha}} \to \sigma - \alpha$ weakly in $L^2(I\times\Omega)$, $(\bar{\bar{\sigma}} - \bar{\bar{\alpha}})\chi \to (\sigma - \alpha)\chi$ weakly in $L^2(I\times\Omega)$. Hence

$$0 \geq \underline{\lim} \int_I\int_\Omega \chi[(\bar{\bar{\sigma}} - \bar{\bar{\alpha}})^2 - z_0^2]\,dx\,dt$$

$$\geq \int_I\int_\Omega \chi[(\sigma - \alpha)^2 - z_0^2]\,dx\,dt$$

and (2.4) follows. Finally, to prove (2.2), we show that for any $\tau \in K_\alpha$ it holds that

$$(2.13) \qquad \int_I (\dot{\sigma} - k\dot{u}_x, \tau - \sigma)dt \geq 0.$$

Define $\theta = (\tau - \alpha)/z_0$. Then $\theta \in L^2(I\times\Omega)$ and $|\theta| \leq 1$ a.e. in $I\times\Omega$. For this θ we can find such $\bar{\bar{\theta}}$ that for given $\varepsilon > 0$ and for sufficiently small h

$$\|\theta - \bar{\bar{\theta}}\|_{L^2(I\times\Omega)} < \varepsilon \ , \ |\bar{\bar{\theta}}| \leq 1 \quad \text{in } I\times\Omega, \qquad \bar{\bar{\theta}} = \sum_{i=1}^{N} \theta_i(t)\bar{\bar{\phi}}_i(x).$$

To see this, let $\rho_\delta *$ be the Friedrichs' mollifier, and consider the function

$$\theta_\delta = \rho_\delta * \theta = \int \rho_\delta(z - y)\theta(y)dy$$

after extending θ by setting $\theta = 0$ outside $I\times\Omega$. Then $|\theta_\delta| \leq 1$ in $I\times\Omega$, and

$$\|\theta - \theta_\delta\|_{L^2(I\times\Omega)} < \frac{\varepsilon}{2}$$

for sufficiently small δ. Since $\theta_\delta \in C^\infty$, the interpolation $\bar{\bar{\theta}}$ of θ by the basis functions $\{\bar{\bar{\phi}}_i\}$ satisfies for sufficiently small h the inequality

$$\|\theta_\delta - \bar{\bar{\theta}}\|_{L^2(I\times\Omega)} < \frac{\varepsilon}{2}.$$

Hence this $\bar{\bar{\theta}}$ is the desired one.

We set $\bar{\bar{\tau}} = \bar{\bar{\alpha}} + z_0\bar{\bar{\theta}}$ in (2.11) and integrate the resulting inequality taking (2.12) into account to get

$$0 \geq \int_0^T (\dot{\bar{\bar{\alpha}}}, \bar{\bar{\alpha}} + z_0\bar{\bar{\theta}} - \bar{\bar{\sigma}})dt$$

$$= \frac{1}{2}\|\bar{\bar{\alpha}}\|^2(T) + z_0\int_0^T (\dot{\bar{\bar{\alpha}}}, \bar{\bar{\theta}})dt - \int_0^T (\dot{\bar{\bar{\alpha}}}, \bar{\bar{\sigma}})dt.$$

As h tends to zero, we have, for a suitable subsequence,

$$\underline{\lim}\ \|\bar{\bar{\alpha}}\|^2(T) \geq \|\alpha\|^2(T),$$

and

$$\int_0^T (\dot{\bar{\bar{\alpha}}},\bar{\bar{\theta}})dt = \int_0^T (\dot{\bar{\bar{\alpha}}},\theta)dt + \int_0^T (\dot{\bar{\bar{\alpha}}},\bar{\bar{\theta}} - \theta)dt \longrightarrow \int_0^T (\dot{\alpha},\theta)dt.$$

Also, since

$$-\int_0^T (\dot{\bar{\bar{\alpha}}},\bar{\bar{\sigma}})dt = \frac{1-\xi}{\xi}\int_0^T (\dot{\bar{\bar{\sigma}}} - k\dot{\hat{u}}_x,\bar{\bar{\sigma}})dt$$

$$= \frac{1-\xi}{2\xi}\|\bar{\bar{\sigma}}\|^2(T) + \frac{k(1-\xi)}{2\xi}\rho(\|\dot{\hat{u}}\|^2(T) - \|\dot{\hat{u}}\|^2(0) - 2\int_0^T(b,\dot{\hat{u}})dt),$$

we have by (2.1) and (2.3)

$$\underline{\lim}\,[-\int_0^T (\dot{\bar{\bar{\alpha}}},\bar{\bar{\sigma}})dt]$$

$$\geq \frac{1-\xi}{2\xi}\|\sigma\|^2(T) + \frac{k(1-\xi)}{2\xi}\rho(\|\dot{u}\|^2(T) - \|\dot{u}\|^2(0) - 2\int_0^T(b,\dot{u})dt)$$

$$= -\int_0^T (\dot{\alpha},\sigma)dt.$$

Hence we have

$$0 \geq \underline{\lim}\int_0^T (\dot{\bar{\bar{\alpha}}},\bar{\bar{\alpha}} + z_0\bar{\bar{\theta}} - \bar{\bar{\sigma}})dt$$

$$\geq \frac{1}{2}\|\alpha\|^2(T) + z_0\int_0^T(\dot{\alpha},\theta)dt - \int_0^T(\dot{\alpha},\sigma)dt$$

$$= \int_0^T(\dot{\alpha},\tau - \sigma)dt$$

from which (2.13) follows.

We next prove that the inequality (2.13) implies that there is a non-negative function λ such that

(2.14) $$\dot{u}_x - \frac{1}{k}\dot{\sigma} = 0 \qquad \text{a.e. } E$$

(2.15) $$\dot{u}_x - \frac{1}{k}\dot{\sigma} = \lambda\frac{\sigma - \alpha}{|\sigma - \alpha|} \qquad \text{a.e. } P.$$

Let θ be an arbitrary constant function on $I\times\Omega$ satisfying $|\theta| \leq 1$, and set

$$\tau = \alpha + \theta z_0.$$

Then evidently $\tau \in K_\alpha$. Let $S_\delta(p)$ be a square centered at $p \in I\times\Omega$ and with side length δ, and define $\tilde{\tau} \in K_\alpha$ by

$$\tilde{\tau} = \begin{cases} \tau & (t,x) \in S_\delta(p) \\ \sigma & \text{otherwise.} \end{cases}$$

We set $\tau = \tilde{\tau}$ in (2.13). Then, since

$$\begin{aligned} 0 &\geq \frac{1}{\delta^2}\int_I (\dot{u}_x - \frac{1}{k}\dot{\sigma}, \tilde{\tau} - \sigma)dt \\ &= \frac{1}{\delta^2}\int_{S_\delta(p)} (\dot{u}_x - \frac{1}{k}\dot{\sigma})(\alpha + \theta z_0 - \sigma)dxdt \\ &\to (\dot{u}_x - \frac{1}{k}\dot{\sigma})\alpha + \theta z_0(\dot{u}_x - \frac{1}{k}\dot{\sigma}) - (\dot{u}_x - \frac{1}{k}\dot{\sigma})\sigma \qquad \text{a.e. } I\times\Omega \end{aligned}$$

as $\delta \to 0$ (Theorem 9.6), we have for any such θ

$$0 \leq (\dot{u}_x - \frac{1}{k}\dot{\sigma})(\sigma - \alpha - \theta z_0) \qquad \text{a.e. } I\times\Omega .$$

Therefore, if $(t,x) \in E$, that is, $|\sigma - \alpha| < z_0$, then

$$\dot{u}_x - \frac{1}{k}\dot{\sigma} = 0$$

and (2.14) holds. If $(t,x) \in P$, then the sign of $\dot{u}_x - \dot{\sigma}/k$ is either 0 or the same as that of $\sigma - \alpha$. Hence (2.15) holds. In this case, λ is given by

$$\text{(2.16)} \qquad \lambda = \frac{\xi}{k(1-\xi)} \frac{\sigma - \alpha}{|\sigma - \alpha|} \dot{\sigma}$$

To prove this we shall show that

$$\text{(2.17)} \qquad \dot{\sigma} - \dot{\alpha} = 0 \qquad \text{a.e. } P.$$

If this is the case, then we have (2.16), since by (2.15)

$$\dot{\sigma} = \dot{\alpha} = \frac{k(1-\xi)}{\xi}(\dot{u}_x - \frac{1}{k}\dot{\sigma}) = \frac{\lambda k(1-\xi)}{\xi} \frac{\sigma - \alpha}{|\sigma - \alpha|} .$$

Now $\dot{\sigma}$ and $\dot{\alpha}$ belong to $L^1(I\times\Omega)$. Thus σ and α are absolutely continuous with respect to t almost everywhere on Ω (Theorem 9.4), and the usual derivative exists for almost all t. Assume now that $(t,x) \in P$ and that σ and α are absolutely continuous with respect to t at this x. Then for almost all t we have

$$0 \geq \lim_{\Delta t \downarrow 0} \frac{[\sigma(t+\Delta t) - \alpha(t+\Delta t)]^2 - [\sigma(t) - \alpha(t)]^2}{\Delta t}$$

$$= 2[\sigma(t) - \alpha(t)][\dot{\sigma}(t) - \dot{\alpha}(t)]$$

$$= \lim_{\Delta t \downarrow 0} \frac{[\sigma(t) - \alpha(t)]^2 - [\sigma(t-\Delta t) - \alpha(t-\Delta t)]^2}{\Delta t} \geq 0.$$

Hence $\dot{\sigma}(t) - \dot{\alpha}(t) = 0$ at this point. In other words, as regards the points of P, $\dot{\sigma} - \dot{\alpha} = 0$ holds for almost all t on almost all the lines which are parallel to the t - axis. Therefore (2.17) holds by (2) of Theorem 9.6. This completes the proof of the theorem.

9.3. Two-dimensional dynamic problem

We modify the original two-dimensional dynamic problem into a slightly weak form. Then we can prove the existence of a solution to this problem as a limit of the solutions of the semidiscrete system (1.5)$\sim$(1.7) of Chapter 5.

Let $D_2^1(\Omega, \Gamma_u)$ be the $W_2^1(\Omega)$ completion of $C^\infty(\bar{\Omega}, \Gamma_u)$ - the set of $C^\infty(\bar{\Omega})$ functions vanishing identically near Γ_u, where Ω is assumed, for simplicity, to be a polygonal region. For $\sigma = (\sigma_{11}, \sigma_{22}, \sigma_{12}) \in L^2(I\times\Omega)$ and $\alpha = (\alpha_{11}, \alpha_{22}, \alpha_{12}) \in L^2(I\times\Omega)$ satisfying

$$f(\sigma - \alpha) \leq z_0 \qquad \text{a.e. } I\times\Omega,$$

we define

$$E = E_{\sigma,\alpha} = \{ (t,x) \in I\times\Omega \ ; \ f(\sigma - \alpha) < z_0 \}$$

$$P = P_{\sigma,\alpha} = \{ (t,x) \in I\times\Omega \ ; \ f(\sigma - \alpha) = z_0 \},$$

where both σ and α are supposed to be bounded at the above (t,x).

DEFINITION. We say that (u, σ, α) is a solution of the two-dimensional dynamic problem if

$$u, \dot{u} \in L^\infty(I; D_2^1(\Omega,\Gamma_u)), \qquad \ddot{u}, \sigma, \alpha, \dot{\sigma}, \dot{\alpha} \in L^\infty(I; L^2(\Omega)),$$

$$u(0,x) = 0, \ \dot{u}(0,x) = a(x), \ \sigma(0,x) = \alpha(0,x) = 0 \qquad \text{a.e. } \Omega ,$$

$$f(\sigma - \alpha) \leq z_0 \qquad \text{a.e. } I\times\Omega,$$

(3.1) $$(\rho\ddot{u}_i,\phi) + \sum_j (\sigma_{ij},\phi_{,j}) = (b_i,\phi) \qquad \text{a.e. } I,$$

$$\text{for all } \phi \in D_2^1(\Omega,\Gamma_u),$$

(3.2) $$\begin{cases} \dot{\sigma} = D\dot{\varepsilon} , \quad (\dot{\alpha} = 0) & \text{a.e. } E \\ \dot{\sigma} = (D - D')\dot{\varepsilon}, \ (\dot{\alpha} = (\sigma - \alpha) \dfrac{\partial f^*\dot{\sigma}}{f}), \quad \partial f\dot{\sigma} \geq 0 & \text{a.e. } P \end{cases}$$

(3.3) $$\dot{\alpha} = \eta S^{-1}(\dot{\varepsilon} - C\dot{\sigma}) \qquad \text{a.e. } I\times\Omega.$$

THEOREM 9.9. The solution of the two-dimensional dynamic problem exists uniquely. It is given by a limit of the solutions of the semidiscrete system (1.5)$\sim$(1.7) of Chapter 5.

PROOF. According to Theorem 5.8, the solution $(\tilde{u}, \tilde{\sigma}, \tilde{\alpha}) \in C_+^1(I)$ of the semidiscrete system satisfies the following relations :

(3.4) $$(\rho\ddot{\tilde{u}}_i,\phi_p) + \sum_j (\tilde{\sigma}_{ij},\phi_{p,j}) = (b_i,\phi_p) \qquad p \in P,$$

(3.5) $(\dot{\tilde{\varepsilon}} - C\dot{\tilde{\sigma}}, \tilde{\tau} - \tilde{\sigma}) \leq 0 \qquad$ for all $\tilde{\tau} \in \tilde{K}$,

(3.6) $\dot{\tilde{\alpha}} = \eta S^{-1}(\dot{\tilde{\varepsilon}} - C\dot{\tilde{\sigma}})$,

where $\tilde{K}$ is defined by

$$\tilde{K} = \{ \tilde{\tau} = \sum_e \tau^e(t)\chi^e;\ \tau^e(t) \in C^1_+(I),\ f(\tau^e - \alpha^e) \leq z_0 \}.$$

Now by the a priori estimates (Theorem 5.5, Theorem 5.7),

$$\left.\begin{array}{l} \|\dot{\tilde{u}}\|,\ \|\tilde{\sigma}\|,\ \|\tilde{\alpha}\| \\ \|\ddot{\tilde{u}}\|,\ \|\dot{\tilde{\sigma}}\|,\ \|\dot{\tilde{\alpha}}\| \end{array}\right\} \quad \text{are uniformly bounded in } L^\infty(I;\ L^2(\Omega)).$$

Also, by Korn's inequality, the boundedness of $\|\tilde{\varepsilon}\|$ and $\|\dot{\tilde{\varepsilon}}\|$ implies that

$$\|\tilde{u}_{i,j}\|,\ \|\dot{\tilde{u}}_{i,j}\| \quad \text{are uniformly bounded in } L^\infty(I;\ L^2(\Omega)).$$

Hence, by Theorem 9.3 there exists $(u, \sigma, \alpha) \in L^\infty(I;\ L^2(\Omega))$ such that for a suitable subsequence

$$\tilde{u},\ \dot{\tilde{u}},\ \ddot{\tilde{u}},\ \tilde{u}_{i,j},\ \dot{\tilde{u}}_{i,j} \rightarrow u,\ \dot{u},\ \ddot{u},\ u_{i,j},\ \dot{u}_{i,j} \qquad \text{weakly* in } L^\infty(I;\ L^2(\Omega))$$

$$\left.\begin{array}{l} \tilde{\sigma},\dot{\tilde{\sigma}} \rightarrow \sigma,\dot{\sigma} \\ \tilde{\alpha},\dot{\tilde{\alpha}} \rightarrow \alpha,\dot{\alpha} \\ \tilde{\varepsilon},\dot{\tilde{\varepsilon}} \rightarrow \varepsilon,\dot{\varepsilon} \end{array}\right\} \quad \text{weakly* in } L^\infty(I;\ L^2(\Omega)).$$

Here we can suppose also that

$$\tilde{u},\ \dot{\tilde{u}} \rightarrow u,\ \dot{u} \qquad \text{weakly* in } L^\infty(I; D^1_2(\Omega,\Gamma_u)) \text{ and weakly in } W^1_2(I\times\Omega).$$

Hence, the initial values of $\tilde{u}$ and $\dot{\tilde{u}}$ converge to those of u and $\dot{u}$ weakly in $L^2(\Omega)$. Since $\tilde{u}(o,x) = 0$ and $\dot{\tilde{u}}(0,x)$ converges to $a(x)$ uniformly on Ω, we have

$$u(0,x) = 0, \quad \dot{u}(0,x) = a(x) \qquad \text{a.e. } \Omega.$$

In the same way, we have $\sigma(0,x) = \alpha(0,x) = 0$ a.e. Ω by the initial conditions of $\tilde{\sigma}$ and $\tilde{\alpha}$.

To prove (3.1), we take an arbitrary $\phi \in C^\infty(\overline{I\times\Omega})$ such that $\phi|_{\Gamma_u} = 0$. Since ϕ can be approximated in $W_2^1(I\times\Omega)$ by the finite element basis $\{\phi_p\}$ as $h \to 0$, we have by (3.4)

$$(3.7) \qquad \int_I [(\rho\ddot{u}_i,\phi) + \sum_j (\sigma_{ij},\phi_{,j})]\,dt = \int_I (b_i,\phi)dt.$$

Let $I_\delta = [t - \delta/2, t + \delta/2] \subset I$ $(\delta > 0)$. For $\bar{\phi} \in D_2^1(\Omega,\Gamma_u)$ set

$$\bar{\psi} = \begin{cases} \bar{\phi} & \text{in } I_\delta \\ 0 & \text{otherwise.} \end{cases}$$

Then (3.7) holds, replacing ϕ by $\bar{\psi}$. To see this, take $\phi_n \in C^\infty(\bar{\Omega}, \Gamma_u)$ which approximates $\bar{\phi}$: $\|\bar{\phi} - \phi_n\|_{W_2^1(\Omega)} \to 0$. Set

$$\psi_n = \begin{cases} \phi_n & \text{in } I_\delta \\ 0 & \text{otherwise.} \end{cases}$$

Then, by approximating ψ_n by C^∞-functions, we see that (3.7) holds for ψ_n and thus

$$\int_{I_\delta} [(\rho\ddot{u}_i,\phi_n) + \sum_j (\sigma_{ij},\phi_{n,j})]\,dt = \int_{I_\delta} (b_i,\phi_n)dt.$$

Since ϕ_n approximates $\bar{\phi}$ in $W_2^1(\Omega)$, this identity holds also for $\bar{\phi}$. Dividing both sides of the above identity (ϕ_n is replaced by $\bar{\phi}$) by δ, and letting $\delta \to 0$, we have by Theorem 9.6

$$(3.8) \qquad (\rho\ddot{u}_i,\bar{\phi}) + \sum_j (\sigma_{ij},\bar{\phi}_{,j}) = (b_i,\bar{\phi})$$

for almost all $t \in I$. Since $D_2^1(\Omega,\Gamma_u)$ is separable, (3.8) holds for all $\bar{\phi}$ in $D_2^1(\Omega,\Gamma_u)$ except t in a set of measure zero.

The equality (3.3) is obtained by going to the limit in (3.6). To prove

(3.2), we first show that

$$(3.9) \qquad \int_I (\dot{\varepsilon} - C\dot{\sigma}, \tau - \sigma)dt \leq 0$$

for all $\tau \in L^\infty(I; L^2(\Omega))$ such that $f(\tau - \alpha) \leq z_0$ a.e. $I\times\Omega$.

The above τ can be expressed as $\tau = \alpha + \theta z_0$ for a suitable $\theta \in L^\infty(I;L^2(\Omega))$ satisfying $f(\theta) \leq 1$ a.e. $I\times\Omega$. We can choose a $\tilde{\theta}$ which satisfies $f(\tilde{\theta}) \leq 1$ and $\|\theta - \tilde{\theta}\|_{L^2(I\times\Omega)} \to 0$, and which has the form

$$\tilde{\theta} = \sum_e \theta^e(t)\chi^e \ ; \quad \theta^e(t) \in C^1_+(I).$$

In fact, let $\rho_\delta *$ be the Friedrichs' mollifier, and set

$$\theta^\delta = \rho_\delta * \theta = \int \rho_\delta(z - y)\theta(y)dy = \int \rho_\delta(y)\theta(z - y)dy.$$

We extended θ by setting $\theta = 0$ outside $I\times\Omega$. Then we have for any $z \in I\times\Omega$

$$\begin{aligned} f^2(\theta^\delta)(z) &= \frac{1}{2}\Big[\int \rho_\delta(y)\theta_{11}(z-y)dy\Big]^2 + \frac{1}{2}\Big[\int \rho_\delta(y)\theta_{22}(z-y)dy\Big]^2 \\ &\quad + \frac{1}{2}\Big[\int \rho_\delta(y)(\theta_{11} - \theta_{22})(z-y)dy\Big]^2 + 3\Big[\int \rho_\delta(y)\theta_{12}(z-y)dy\Big]^2 \\ &\leq \int \rho_\delta(z-y)\Big[\frac{1}{2}\theta_{11}^2(y) + \frac{1}{2}\theta_{22}^2(y) + \frac{1}{2}(\theta_{11} - \theta_{22})^2(y) + 3\theta_{12}^2(y)\Big]dy \\ &= \int \rho_\delta(z-y)f^2(\theta(y))dy \leq 1. \end{aligned}$$

Let $\tilde{\theta}$ be the interpolation of θ^δ by the basis function $\{\chi^e\}$. Then, since $\theta^\delta \in C^\infty(\overline{I\times\Omega})$, we have $\|\theta^\delta - \tilde{\theta}\|_{L^2(I\times\Omega)} \to 0$ as $h \to 0$. Hence this $\tilde{\theta}$ is the desired one since $\theta^\delta \to \theta$ in $L^2(I\times\Omega)$, as is well known.

Now substituting $\tilde{\tau} = \tilde{\alpha} + z_0\tilde{\theta}$ into (3.5) and integrating the resulting inequality with respect to t, we have

$$0 \geq \frac{1}{2\eta}\|\tilde{\alpha}\|_S^2(T) + \frac{z_0}{\eta}\int_0^T (S\dot{\tilde{\alpha}}, \tilde{\theta})dt - \int_0^T (\dot{\tilde{\varepsilon}} - C\dot{\tilde{\sigma}}, \tilde{\sigma})dt.$$

The last term of the right side is written by (3.4) as

$$\frac{1}{2}\rho[\|\dot{\tilde{u}}\|^2(T) - \|\dot{\tilde{u}}\|^2(0)] + \frac{1}{2}\|\tilde{\sigma}\|_C^2(T) - \int_0^T (b,\dot{\tilde{u}})dt.$$

Since $\dot{\tilde{u}}(T)$, $\tilde{\sigma}(T)$ and $\tilde{\alpha}(T)$ can be regarded as weakly converging sequences in $L^2(\Omega)$, then $\dot{u}(T), \sigma(T)$ and $\alpha(T)$ exist as their respective limits. Also, since

$$\int_I (S\dot{\tilde{\alpha}},\tilde{\theta})dt = \int_I [(S\dot{\tilde{\alpha}},\theta) + (S\dot{\tilde{\alpha}},\tilde{\theta} - \theta)]\, dt$$

$$\to \int_I (S\dot{\alpha},\theta)dt,$$

we have the desired inequality

$$0 \geq \frac{1}{2\eta}\|\alpha\|_S^2(T) + \frac{z_0}{\eta}\int_I (S\dot{\alpha},\theta)dt$$

$$+ \frac{1}{2}\rho[\|\dot{u}\|^2(T) - \|\dot{u}\|^2(0)] + \frac{1}{2}\|\sigma\|_C^2(T) - \int_I (b,\dot{u})dt$$

$$= \int_I (\dot{\varepsilon} - C\dot{\sigma}, \tau - \sigma)dt.$$

We next prove that (3.9) implies that there is a nonnegative measurable function $\lambda(t,x)$ such that

(3.10) $$\dot{\varepsilon} - C\dot{\sigma} = 0 \qquad \text{a.e. } E$$

(3.11) $$\dot{\varepsilon} - C\dot{\sigma} = \lambda\, \partial f \qquad \text{a.e. } P.$$

Take $\theta \in R^3$ satisfying $f(\theta) \leq 1$, and set $\tau = \alpha + \theta z_0$ in (3.9). Then, by the same argument as in the rod problem, we have

(3.12) $$0 \leq (\dot{\varepsilon} - C\dot{\sigma})^*(\sigma - \alpha - \theta z_0) \qquad \text{a.e. } I\times\Omega.$$

Hence, if $(t,x) \in E$ (that is if $f(\sigma-\alpha) < z_0$), then clearly (3.10) must holds. On the other hand, if $(t,x) \in P$, then (3.12) implies that the vector $\dot{\varepsilon} - C\dot{\sigma}$ is either zero or parallel to the outward normal to the surface $f(\sigma - \alpha) = z_0$ at the stress point σ, and hence (3.11) must hold. We shall show below that

(3.13) $$\lambda = \frac{1}{\eta}\partial f^*\dot{\sigma}\ .$$

If this is the case, the relation $\dot{\sigma} = (D - D')\dot{\varepsilon}$ a.e. P is evident. To prove (3.13) it suffices to show

$$(3.14) \qquad \partial f^*(\dot{\sigma} - \dot{\alpha}) = 0 \qquad \text{a.e. } P,$$

since if this relation holds, we can substitute (3.3) and (3.11) into (3.14) to get

$$0 = \partial f^*(\dot{\sigma} - \eta S^{-1}\lambda\partial f\,) = \partial f^*\dot{\sigma} - \lambda\eta\,.$$

Note that $S^{-1}\partial f = (\sigma - \alpha)/f$ holds identically. The proof of (3.14) is essentially the same as in the rod problem. Assume that $(t,x) \in P$, and that σ and α are absolutely continuous on t at this x. We then have for almost all such t

$$0 \geq \lim_{\Delta t \to +0} [\, f^2(t + \Delta t) - f^2(t)\,]/\Delta t \qquad (\, f(t) = f(\sigma(t)) \text{ etc. })$$

$$= 2f\partial f^*(\dot{\sigma} - \dot{\alpha})$$

$$= \lim_{\Delta t \to +0} [\, f^2(t) - f^2(t - \Delta t)\,]\,/\Delta t \geq 0.$$

Hence (3.14) holds. This proves (3.2). Finally the proof of $f(\sigma - \alpha) \leq z_0$ a.e. $I\times\Omega$. This relation is proved by noticing that $f(\sigma)$ can be regarded as a norm in R^3, which is equivalent to the Euclidian norm. But we shall make use of the convexity of the function f. Define the set

$$B_f = \{\, (\tau,\zeta) \in R^3\times R^3 \;;\; f(\tau - \zeta) \leq z_0\}\,.$$

Let $P_f : R^3\times R^3 \to B_f$ be the projection in Euclidian space R^6 and consider the inner product of vectors

$$(\,(\sigma,\alpha) - P_f(\sigma,\alpha),\ (\tau,\zeta) - P_f(\sigma,\alpha))_{R^6} \qquad (\tau,\zeta)\in B_f\,.$$

Since this value is nonpositive for a.e. $I\times\Omega$, we have

$$\iint_{I\times\Omega}((\sigma,\alpha) - P_f(\sigma,\alpha), (\tau,\zeta) - P_f(\sigma,\alpha))_{R^6}\, dxdt \leq 0.$$

Here we can set $(\tau,\zeta) = (\tilde{\sigma},\tilde{\alpha})$. Passing to the limit with respect to h, we have

$$\iint_{I\times\Omega}\| (\sigma,\alpha) - P_f(\sigma,\alpha)\|_{R^6}\, dxdt = 0,$$

from which (3.15) follows. The uniqueness is readily proved.

9.4. Isotropic hardening problem

We re-define the isotropic hardening problem as follows : Let G be the function defined by (6.13) of Chapter 2, and set

$$B_H = \{(\tau,\zeta) \in R^3\times(-\delta,\infty) \ ; \ f(\tau) \leq H(\zeta)\}$$
$$B_G = \{(\tau,\zeta) \in R^3\times(-\delta^*,\infty) \ ; \ f(\tau) \leq G(\zeta)\}$$
$$K_H = \{(\tau,\zeta) \in [L^\infty(I; L^2(\Omega))]^3 \times L^\infty(I; L^2(\Omega)) \ ; \ (\tau,\zeta) \in B_H \ \text{ a.e. } I\times\Omega\}$$
$$K_G = \{(\tau,\zeta) \in [L^\infty(I; L^2(\Omega))]^3 \times L^\infty(I; L^2(\Omega)) \ ; \ (\tau,\zeta) \in B_G \ \text{ a.e. } I\times\Omega\}.$$

Note that B_G and B_H are closed and convex in R^4. Let $(\sigma, \bar{\varepsilon}^p) \in K_H$, and introduce the sets

$$E = E_{\sigma,\bar{\varepsilon}^p} = \{ (t,x) \in I\times\Omega \ ; \ f(\sigma) < H(\bar{\varepsilon}^p)\}$$
$$P = P_{\sigma,\bar{\varepsilon}^p} = \{ (t,x) \in I\times\Omega \ ; \ f(\sigma) = H(\bar{\varepsilon}^p)\}$$

DEFINITION. We say that $(u, \sigma, \bar{\varepsilon}^p)$ is a solution of the two-dimensional dynamic problem with the isotropic hardening rule if

$$(4.1)\qquad \begin{cases} u,\ \dot{u} \in L^\infty(I; D^1_2(\Omega,\Gamma_u)), \quad \ddot{u},\ \sigma,\ \dot{\sigma},\ \bar{\varepsilon}^p,\ \dot{\bar{\varepsilon}}^p \in L^\infty(I; L^2(\Omega)) \\ u(0,x) = 0,\ \dot{u}(0,x) = a(x),\ \sigma(0,x) = \bar{\varepsilon}^p(0,x) = 0,\ \varepsilon = \varepsilon(u), \\ f(\sigma) \leq H(\bar{\varepsilon}^p) \qquad \text{a.e. } I\times\Omega, \end{cases}$$

(4.2) $$(\rho\ddot{u}_i,\phi) + \sum_{j=1}^{2} (\sigma_{ij},\phi_{,j}) = (b_i,\phi) \qquad \text{a.e. } I$$

for all $\phi \in D_2^1(\Omega, \Gamma_u)$, and

(4.3) $$\begin{cases} \dot{\sigma} = D\dot{\varepsilon} & \text{a.e. } E \\ \dot{\sigma} = (D - D')\dot{\varepsilon} & \text{a.e. } P, \end{cases}$$

where

$$D' = \frac{D\partial f \partial f^* D}{H'(\bar{\varepsilon}^p) + \partial f^* D \partial f},$$

(4.4) $$\bar{\varepsilon}^p = \int_0^t \frac{\sigma^* \dot{\varepsilon}^p}{f(\sigma)}\, dt \quad (\dot{\bar{\varepsilon}}^p \geq 0) \quad (\dot{\varepsilon}^p = \dot{\varepsilon} - C\dot{\sigma}\ ,\ \dot{\varepsilon} = \varepsilon(\dot{u})\ ).$$

THEOREM 9.10. The two-dimensional dynamic problem with the isotropic hardening rule has a unique solution, which is given by a limit of the solutions of the semidiscrete system.

PROOF. We shall use the suffix h to denote the solution of the semi-discrete system. Now it is evident that there is (u, σ) which is a limit of the solutions (u_h, σ_h) of the semidiscrete system and satisfy (4.1) and (4.2), except the conditions including $\bar{\varepsilon}^p$. Also there exists a $\hat{\varepsilon}^p \in L^\infty(I;L^2(\Omega))$ as a limit of $\hat{\varepsilon}_h^p$ - the hardening parameter of the semidiscrete system. Since $\hat{\varepsilon}_h^p$ is nonnegative, this $\hat{\varepsilon}^p$ is also nonnegative a.e. $I\times\Omega$. Furthermore, it holds that

(4.5) $$f(\sigma) \leq G(\hat{\varepsilon}^p) \qquad \text{a.e. } I\times\Omega.$$

To show this, let $P_G : R^3 \times R^1 \to B_G$ be the projection, and consider the inequality which holds for the approximate solutions

$$0 \geq \iint_{I\times\Omega} ((\sigma,\hat{\varepsilon}^p) - P_G(\sigma,\hat{\varepsilon}^p),\ (\sigma_h,\hat{\varepsilon}_h^p) - P_G(\sigma,\hat{\varepsilon}^p)\ )_{R^4}\, dxdt.$$

Inequality (4.5) is then obtained by going to the limit in this inequality.

To prove (4.3) we first note that for all $(\tau,\zeta) \in K_G$

$$(4.6) \qquad 0 \geq \iint_{I\times\Omega} [(\dot{\varepsilon} - C\dot{\sigma}, \tau - \sigma) - \dot{\hat{\varepsilon}}^p(\zeta - \hat{\varepsilon}^p)]\, dxdt$$

which is obtained by going to the limit in the inequality

$$0 \geq \iint_{I\times\Omega} [(\dot{\varepsilon}_h - C\dot{\sigma}_h, \tau - \sigma_h) - \dot{\hat{\varepsilon}}^p_h(\zeta - \hat{\varepsilon}^p_h)]\, dxdt$$

The last inequality itself is a direct consequence of Theorem 5.13.

Let $S_\delta(p)$ $(p \in I\times\Omega)$ be a cube with side length δ, centered at p. Let $(\tau,\zeta) \in B_G$, and set

$$\tau_* = \begin{cases} \tau & \text{in } S_\delta(p) \\ \sigma & \text{in } I\times\Omega - S_\delta(p) \end{cases} \qquad \zeta_* = \begin{cases} \zeta & \text{in } S_\delta(p) \\ \hat{\varepsilon}^p & \text{in } I\times\Omega - S_\delta(p). \end{cases}$$

Then clearly $(\tau_*,\zeta_*) \in K_G$. Hence we have by (4.6)

$$0 \geq \frac{1}{\text{measure}[S_\delta(p)]} \int_{S_\delta(p)} [(\dot{\varepsilon} - C\dot{\sigma}, \tau - \sigma) - \dot{\hat{\varepsilon}}^p(\zeta - \hat{\varepsilon}^p)]\, dxdt.$$

Letting $\delta \to 0$, we have by Theorem 9.6

$$(4.7) \qquad (\dot{\varepsilon} - C\dot{\sigma}, \tau - \sigma)_{R^3} - \dot{\hat{\varepsilon}}^p(\zeta - \hat{\varepsilon}^p) \leq 0 \qquad \text{a.e. } I\times\Omega$$

for all $(\tau,\zeta) \in B_G$. Hence if $f(\sigma)<G(\hat{\varepsilon}^p)$ at (t,x) it is clear that

$$\dot{\varepsilon} - C\dot{\sigma} = 0, \quad -\dot{\hat{\varepsilon}}^p = 0.$$

Also, since $\sigma(0,x) = \hat{\varepsilon}^p(0,x) = 0$, $\hat{\varepsilon}^p(t,x) = 0$ holds until the point $(\sigma,0)$ in B_G reaches its boundary. On the other hand, if $f(\sigma)=G(\hat{\varepsilon}^p)$ at (t,x), then (4.7) implies that there is a nonnegative function $\lambda = \lambda(t,x)$ satisfying for almost all such points

$$(4.8)\qquad \begin{pmatrix} \dot{\varepsilon} - C\dot{\sigma} \\ -\dot{\hat{\varepsilon}}^p \end{pmatrix} = \lambda \begin{pmatrix} \partial f \\ -G' \end{pmatrix} \qquad \lambda \geq 0.$$

We next define $\bar{\varepsilon}^p$ by the inverse transformation of

$$(4.9)\qquad \hat{\varepsilon}^p = \int_0^{\bar{\varepsilon}^p} \sqrt{H'(\lambda)}\, d\lambda \,.$$

Then clearly $\bar{\varepsilon}^p$ and $\dot{\bar{\varepsilon}}^p$ belong to $L^\infty(I;\ L^2(\Omega))$ and $G(\hat{\varepsilon}^p) = H(\bar{\varepsilon}^p)$ holds by the definition of G. Also the function λ in (4.8) is given by

$$(4.10)\qquad \lambda = \dot{\bar{\varepsilon}}^p.$$

To show this, we first note that $\bar{\varepsilon}^p(t,x)$ is absolutely continuous on t for almost all x since $\hat{\varepsilon}^p(t,x)$ is . Hence we have by (4.8) and (4.9)

$$G'(\hat{\varepsilon}^p)\lambda = \dot{\hat{\varepsilon}}^p = \sqrt{H'(\bar{\varepsilon}^p)}\,\dot{\bar{\varepsilon}}^p = G'(\hat{\varepsilon}^p)\dot{\bar{\varepsilon}}^p$$

from which (4.10) follows. We next show that

$$(4.11)\qquad \partial f^*\dot{\sigma} - H'(\bar{\varepsilon}^p)\dot{\bar{\varepsilon}}^p = 0 \qquad \text{a.e. } P.$$

Take $(t,x) \in P$; that is, assume that $f(\sigma) = G(\hat{\varepsilon}^p) = H(\bar{\varepsilon}^p)$ at this point. Since σ and $\bar{\varepsilon}^p$ are absolutely continuous on t for almost all $x \in \Omega$, we have at such x and for almost all such t

$$\begin{aligned} 0 &\geq \lim_{\Delta t \to +0} [\, f(t+\Delta t) - H(t+\Delta t) - (f(t) - H(t))] \\ &= \partial f^*\dot{\sigma} - H'(\bar{\varepsilon}^p)\dot{\bar{\varepsilon}}^p \qquad (\, f(t) = f(\sigma(t)) \text{ etc. }) \\ &= \lim_{\Delta t \to +0} [\, f(t) - H(t) - (f(t-\Delta t) - H(t-\Delta t))] \geq 0. \end{aligned}$$

This proves (4.11). Therefore, we finally have by (4.8)

$$(4.12)\qquad \dot{\varepsilon} - C\dot{\sigma} = \partial f\dot{\bar{\varepsilon}}^p = \partial f\,\frac{\partial f^*\dot{\sigma}}{H'(\bar{\varepsilon}^p)} \qquad \text{a.e. } P$$

from which follows

$$\dot{\sigma} = (D - D')\dot{\varepsilon} \qquad \text{a.e. } P, \quad \text{where} \qquad D' = \frac{D\partial f\partial f^* D}{H' + \partial f^* D\partial f}$$

This proves (4.3). Let us next prove that $\bar{\varepsilon}^p$ is given by (4.4). By (4.12) we have

$$\dot{\varepsilon}^p = \partial f \, \frac{\partial f^* \dot{\sigma}}{H'(\bar{\varepsilon}^p)} \qquad \text{a.e. } P.$$

Therefore, taking into account (4.8), we have a.e. P

$$\dot{\hat{\varepsilon}}^p = G'(\hat{\varepsilon}^p)\lambda = G'(\hat{\varepsilon}^p) \, \frac{\partial f^* \dot{\sigma}}{H'(\bar{\varepsilon}^p)} = G'(\hat{\varepsilon}^p) \, \frac{\sigma^* \dot{\varepsilon}^p}{f(\sigma)}$$

These equalities hold a.e. on $I \times \Omega$, since $\dot{\varepsilon}^p = \dot{\hat{\varepsilon}}^p = 0$ a.e. on E. Hence we have

$$\bar{\varepsilon}^p = \int_0^t \dot{\bar{\varepsilon}}^p \, dt = \int_0^t \frac{\dot{\hat{\varepsilon}}^p}{G'(\hat{\varepsilon}^p)} \, dt = \int_0^t \frac{\sigma^* \dot{\varepsilon}^p}{f(\sigma)} \, dt \, ,$$

which proves (4.4).

To prove the uniqueness, let $(u, \sigma, \bar{\varepsilon}^p)$ be a solution. By transformation (4.9) we introduce a new parameter $\hat{\varepsilon}^p$. Since $\dot{\hat{\varepsilon}}^p = \sqrt{H'}\dot{\bar{\varepsilon}}^p = G'\dot{\bar{\varepsilon}}^p$, we have the identity

$$\begin{pmatrix} \dot{\varepsilon} - C\dot{\sigma} \\ -\dot{\hat{\varepsilon}}^p \end{pmatrix} = \begin{pmatrix} \partial f \\ -G'(\hat{\varepsilon}^p) \end{pmatrix} \dot{\bar{\varepsilon}}^p \qquad \text{a.e. } I \times \Omega$$

Hence the inequality

$$\iint_{I \times \Omega} [(\dot{\varepsilon} - C\dot{\sigma}, \tau - \sigma)_{R^3} - \dot{\hat{\varepsilon}}^p(\zeta - \hat{\varepsilon}^p)] \, dx dt \le 0$$

holds for all $(\tau, \zeta) \in K_G$. Thus the uniqueness of $(u, \sigma, \hat{\varepsilon}^p)$ is readily proved and $(u, \sigma, \bar{\varepsilon}^p)$ is the unique solution.

9.5. Two-dimensional quasi-static problem

It is now easy to show that the solutions of the semidiscrete quasi-static system converge and that its limit is a solution of the fully continuous problem. The following theorems are the quasi-static versions of Theorems 9.8 and 9.9, and proved in the same way, since we already have the necessary a priori estimates.

THEOREM 9.11. There exists a unique solution (u, σ, α) of the following quasi-static problem with the kinematic hardening rule :

$$(5.1)\quad \begin{cases} u,\dot{u} \in L^{\infty}(I;\ D_2^1(\Omega,\Gamma_u)), & \sigma,\dot{\sigma},\alpha,\dot{\alpha} \in L^{\infty}(I;\ L^2(\Omega)) \\ u(0,x) = \sigma(0,x) = \alpha(0,x) = 0, & \varepsilon = \varepsilon(u) \\ f(\sigma - \alpha) \leq z_0 & \text{a.e. } I\times\Omega \end{cases}$$

$$(5.2)\quad \sum_{j=1}^{2} (\sigma_{ij},\phi_{,j}) = (b_i,\phi) \quad \text{a.e. } I$$

$$\text{for all } \phi \in D_2^1(\Omega,\Gamma_u)$$

$$(5.3)\quad \begin{cases} \dot{\sigma} = D\dot{\varepsilon}, \quad (\dot{\alpha} = 0) & \text{a.e. } E \\ \dot{\sigma} = (D - D')\dot{\varepsilon}, \quad (\dot{\alpha} = (\sigma - \alpha)\dfrac{\partial f^*\dot{\sigma}}{f(\sigma - \alpha)}), \quad \partial f^*\dot{\sigma} \geq 0 & \text{a.e. } P \end{cases}$$

$$(5.4)\quad \dot{\alpha} = \eta S^{-1}(\dot{\varepsilon} - C\dot{\sigma}) \quad \text{a.e. } I\times\Omega ,$$

where E and P are those defined on page 190.

THEOREM 9.12. There exists a unique solution $(u, \sigma, \bar{\varepsilon}^p)$ of the following quasi-static problem with the isotropic hardening rule :

$$(5.5)\quad \begin{cases} u,\dot{u} \in L^{\infty}(I;\ D_2^1(\Omega,\Gamma_u)), & \sigma,\dot{\sigma},\bar{\varepsilon}^p,\dot{\bar{\varepsilon}}^p \in L^{\infty}(I;\ L^2(\Omega)) \\ u(0,x) = \sigma(0,x) = \bar{\varepsilon}^p(0,x) = 0, & \varepsilon = \varepsilon(u) \\ f(\sigma) \leq H(\bar{\varepsilon}^p) & \text{a.e. } I\times\Omega , \end{cases}$$

$$(5.6) \qquad \sum_{j=1}^{2} (\sigma_{ij}, \phi_{,j}) = (b_i, \phi) \qquad \text{a.e. } I$$

$$\text{for all } \phi \in D_2^1(\Omega, \Gamma_u),$$

$$(5.7) \qquad \begin{cases} \dot{\sigma} = D\dot{\varepsilon} & \text{a.e. } E \\ \dot{\sigma} = (D - D')\dot{\varepsilon} & \text{a.e. } P, \end{cases}$$

where

$$D' = \frac{D\partial f \partial f^* D}{H'(\bar{\varepsilon}^p) + \partial f^* D \partial f}$$

$$(5.8) \qquad \bar{\varepsilon}^p = \int_0^t \frac{\sigma^* \dot{\varepsilon}^p}{f(\sigma)}\, dt \qquad (\dot{\bar{\varepsilon}}^p \geq 0) \qquad (\dot{\varepsilon}^p = \dot{\varepsilon} - C\dot{\sigma}),$$

and where E and P are those defined on page 196 .

REMARK. In formulating the quasi-static problems we added the condition " or $f(\sigma - \alpha) = z_0$ and $\partial f^* \dot{\sigma} < 0$," or " or $f(\sigma) = H(\bar{\varepsilon}^p)$ and $\partial f^* \dot{\sigma} < 0$," which represent the unloading. In the above results, these conditions disappear since such cases have been discarded because of measure zero. Note that the derivative $\dot{\sigma}$ appearing in (5.7) is the " ordinal derivative, " while the original one is the " right derivative."

9.6. Strong convergence and error estimates of the finite element solutions

We have proved the convergence of the finite element solutions to the exact solution in a certain topology. In this section, we shall prove that these approximate solutions converge in the energy norm, too.

We shall use the following abbreviations : The first D and QS denote " dynamic " and " quasi-static." The second S and D denote " semidiscrete " and " discrete." Also, (k) and (i) denote " kinematic " and " isotropic. " Hence, D·S(k) and QS·D(i), for example, denote respectively " dynamic semi-discrete problem with the kinematic hardening rule " and " quasi-static discrete problem with the isotropic hardening rule. "

THEOREM 9.13. Let (u,σ,α) and $(\tilde{u},\tilde{\sigma},\tilde{\alpha})$ be the solutions of D(k) and D·S(k) respectively. Set

$$E(t)^2 = \rho\|\dot{u} - \dot{\tilde{u}}\|^2 + \|\sigma - \tilde{\sigma}\|_C^2 + \frac{1}{\eta}\|\alpha - \tilde{\alpha}\|_S^2$$

$$\delta_0 = \|a(x) - \dot{\tilde{u}}(0,x)\|_{L^2(\Omega)} \qquad \delta_1 = \|\dot{u} - \tilde{v}\|_{L^2(I\times\Omega)}$$

$$\delta_2 = \|\dot{\varepsilon} - \tilde{\varepsilon}^*\|_{L^2(I\times\Omega)} \qquad (\tilde{\varepsilon}^* = \varepsilon(\tilde{v})),$$

where $\tilde{v} = \sum_p v_p(t)\phi_p$ is the best approximation of $\dot{u}$ in $W_2^1(I\times\Omega)$ norm. Then we have the error estimate

$$E(t) \le C(\delta_0 + \sqrt{\delta_1} + \delta_2) \qquad t \in I. \tag{6.1}$$

PROOF. We start from a result of Theorem 5.8 :

$$(\rho\ddot{\tilde{u}}_i,\phi_p) + \sum_j (\tilde{\sigma}_{ij},\phi_{,p}) = (b_i,\phi_p) \qquad p \in P \tag{6.2}$$

$$\int_0^t (\dot{\tilde{\varepsilon}} - C\dot{\tilde{\sigma}},\tilde{\tau} - \tilde{\sigma})dt \le 0 \qquad \text{for all } \tilde{\tau} \in \tilde{K} \quad (t \in I) \tag{6.3}$$

$$\dot{\tilde{\alpha}} = \eta S^{-1}(\dot{\tilde{\varepsilon}} - C\dot{\tilde{\sigma}}), \tag{6.4}$$

where $\tilde{K} = \{\tau = \sum_e \tau^e(t)\chi^e \,;\, \tau^e \in C_+^1(I),\ f(\tau - \alpha) \le z_0\}$. As seen in the proof of Theorem 9.9 or in its proof, these relations hold for (u,σ,α), replacing ϕ_p and $\tilde{\tau}$ respectively by $\phi \in D_2^1(\Omega,\Gamma_u)$ and $\tau \in L^\infty(I;L^2(\Omega))$ satisfying $f(\tau - \alpha) \le z_0$

a.e. on $I\times\Omega$. It is also easy to see that inequality (6.3) holds for all functions of the form

$$\tau = \tilde{\alpha} + \theta z_0 \qquad (\theta \in L^\infty(I; L^2(\Omega)) \ ; \ f(\theta) \le 1 \qquad \text{a.e. } I\times\Omega \ ).$$

Hence we have the two inequalities

$$\int_0^t (\dot{\tilde{\varepsilon}} - C\dot{\tilde{\sigma}}, \tilde{\alpha} + \frac{\sigma - \alpha}{z_0} z_0 - \tilde{\sigma})dt \le 0$$

$$\int_0^t (\dot{\varepsilon} - C\dot{\sigma}, \alpha + \frac{\tilde{\sigma} - \tilde{\alpha}}{z_0} z_0 - \sigma)dt \le 0.$$

Adding these inequalities we have, by taking into account (6.4),

$$(6.5) \qquad 0 \ge \frac{1}{\eta}\int_0^t (S[\dot{\alpha} - \dot{\tilde{\alpha}}], \alpha - \tilde{\alpha})dt + \int_0^t (C[\dot{\sigma} - \dot{\tilde{\sigma}}], \sigma - \tilde{\sigma})dt$$
$$- \int_0^t (\dot{\varepsilon} - \dot{\tilde{\varepsilon}}, \sigma - \tilde{\sigma})dt.$$

On the other hand, by (6.2) we have

$$0 = (\rho[\ddot{\tilde{u}} - \ddot{u}], \dot{\tilde{u}} - \tilde{v}) + (\tilde{\sigma} - \sigma, \dot{\tilde{\varepsilon}} - \tilde{\varepsilon}^*)$$
$$= (\rho[\ddot{\tilde{u}} - \ddot{u}], \dot{\tilde{u}} - \dot{u}) + (\tilde{\sigma} - \sigma, \dot{\tilde{\varepsilon}} - \dot{\varepsilon})$$
$$+ (\rho[\ddot{\tilde{u}} - \ddot{u}], \dot{u} - \tilde{v}) + (\tilde{\sigma} - \sigma, \dot{\varepsilon} - \tilde{\varepsilon}^*).$$

Subsituting this into (6.5) we have

$$E(t)^2 \le C(\delta_0^2 + \delta_1 + \delta_2 \int_0^t E(s)ds),$$

from which (6.1) follows.

THEOREM 9.14. Let $(u,\sigma,\bar{\varepsilon}^p)$ and $(u_h,\sigma_h, \bar{\varepsilon}_h^p)$ be the solutions of D(i) and D·S(i), respectively. Then we have the error estimate

$$\rho\| \dot{u} - \dot{u}_h \| + \|\sigma - \sigma_h\| + \|\bar{\varepsilon}^p - \bar{\varepsilon}_h^p\| \le C(\delta_0 + \sqrt{\delta_1} + \delta_2) \qquad t \in I,$$

where δ_0, δ_1 and δ_2 are those defined in Theorem 9.13.

PROOF. In this case, we have the inequalities

$$(6.6) \qquad 0 \geq \int_0^t\int_\Omega [(\dot{\varepsilon} - C\dot{\sigma}, \tau - \sigma) - \dot{\hat{\varepsilon}}^p(\zeta - \hat{\varepsilon}^p)]\,dxdt$$

$$(6.7) \qquad 0 \geq \int_0^t\int_\Omega [(\dot{\varepsilon}_h - C\dot{\sigma}_h, \tau - \sigma_h) - \dot{\hat{\varepsilon}}^p_h(\zeta - \hat{\varepsilon}^p_h)]\,dxdt$$

for all $(\tau,\zeta)\in K_G$. Hence, setting $(\tau,\zeta) = (\sigma_h, \hat{\varepsilon}^p_h)$ in (6.6) and $(\tau,\zeta) = (\sigma, \hat{\varepsilon}^p)$ in (6.7) and adding the resulting inequalities, we have

$$0 \geq \int_0^t\int_\Omega [(\dot{\varepsilon} - \dot{\varepsilon}_h - C(\dot{\sigma} - \dot{\sigma}_h), \sigma_h - \sigma) - (\dot{\hat{\varepsilon}}^p - \dot{\hat{\varepsilon}}^p_h)(\hat{\varepsilon}^p_h - \hat{\varepsilon}^p)]\,dxdt$$
$$= \int_0^t\int_\Omega (C[\dot{\sigma} - \dot{\sigma}_h], \sigma - \sigma_h)dxdt + \int_0^t\int_\Omega (\dot{\hat{\varepsilon}}^p - \dot{\hat{\varepsilon}}^p_h)(\hat{\varepsilon}^p - \hat{\varepsilon}^p_h)dxdt$$
$$- \int_0^t\int_\Omega (\dot{\varepsilon} - \dot{\varepsilon}_h, \sigma - \sigma_h)dxdt.$$

Thus the desired estimate is obtained the same way as before.

REMARK. $\delta_0 = O(h)$ if $a(x)$ satisfies suitable (but natural) conditions. Also, since the piecewise linear finite elements are dense in W_2^1, δ_1 and δ_2 tends to zero as $h \to 0$, although the order of convergence is not clear.

The essential part of the convergence proof for dynamic problems is applicable to the quasi-static problems. We proved in Chapter 8 that the discrete solutions converge to the solution of the semidiscrete problem, and we obtained the order of this convergence. The following theorems, then are the final results in the quasi-static problems treated in this monograph.

THEOREM 9.15. Let (u,σ,α), $(\tilde{u},\tilde{\sigma},\tilde{\alpha})$ and (u^n,σ^n,α^n) ($0 \leq n \leq T/\Delta t$) be the solutions of QS(k), QS·S(k) and QS·D(k), respectively. Then we have the following estimates in I :

$$\|\sigma - \tilde{\sigma}\|_C + \frac{1}{\eta}\|\alpha - \tilde{\alpha}\|_S \leq C\delta_2$$

$$\|\sigma(n\Delta t) - \sigma^n\|_C + \frac{1}{\eta}\|\alpha(n\Delta t) - \alpha^n\|_S \leq C(\Delta t^{\frac{1}{4}} + \delta_2).$$

THEOREM 9.16. Let $(u, \sigma, \bar{\varepsilon}^p)$, $(u_h, \sigma_h, \bar{\varepsilon}_h^p)$ and $(u^n, \sigma^n, \bar{\varepsilon}_n^p)$ $(0 \leq n \leq T/\Delta t)$ be the solutions of QS(i), QS·S(i) and QS·D(i), respectively. Then we have the following estimates in I :

$$\|\sigma - \sigma_h\|_C + \|\hat{\varepsilon}^p - \hat{\varepsilon}_h^p\| \leq C\delta_2$$

$$\|\sigma(n\Delta t) - \sigma^n\|_C + \|\hat{\varepsilon}^p(n\Delta t) - \hat{\varepsilon}_n^p\| \leq C(\sqrt{\frac{\Delta t}{h}} + \delta_2),$$

where

$$\hat{\varepsilon}^p = \int_0^{\bar{\varepsilon}^p} \sqrt{H'(\lambda)}d\lambda , \quad \hat{\varepsilon}_n^p = \sum_{i=0}^{n-1} \sqrt{H_i'}\, D_t\bar{\varepsilon}_i^p .$$

CHAPTER 10

INTRODUCTION TO AN ELASTIC-PLASTIC PROBLEM WITH GEOMETRICAL NONLINEARITY

10.1. A nonlinear beam problem

All problems discussed in the preceding chapters are based on the assumption that the deflection is small, so the strains are given by a linear combination of the first derivatives of the displacements. If we admit large deflection, however, this linearity does not hold. Also, the space dimension increases, and this causes an additional difficulty in both mathematical and numerical treatments. In this chapter we consider a vibration problem with these nonlinearities and show that the solution of this problem is obtained by a limit of semidiscrete solutions derived from a finite element approximation.

Consider a straight thin beam which, for simplicity's sake, is assumed to be clamped at the ends. We shall first introduce a classical formulation of the elastic-plastic vibration of this beam which admit a large deflection. We use the following notations :

x, y, z : coodinates of particle before deformation (Fig. 16).
The beam is assumed to occupy the region $V_0 = \Omega \times Y \times Z$ in (x,y,x)-space, where $\Omega = [0,L]$, $Y = [o,b]$ and $Z = [-a/2,a/2]$ $(a > 0)$.

u, w : displacemets of the middle plane (i.e., $z = 0$) of the beam in x and z directions, respectively, which are assumed to be constant in y direction.

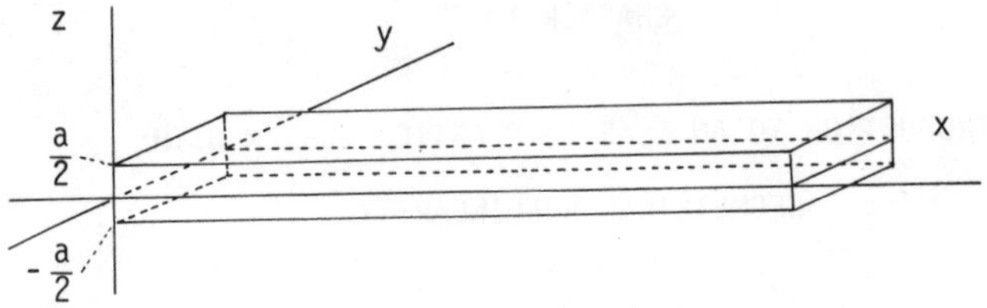

Fig. 16 The coordinate axes for a beam

Let U and W be the displacements of the beam in x and z directions, respectively. We use the following approximations of the displacements and the strain :

(1.1)
$$U = u - z\frac{\partial w}{\partial x}$$
$$W = w$$

(1.2)
$$\varepsilon = \frac{\partial u}{\partial x} + \frac{1}{2}\left(\frac{\partial W}{\partial x}\right)^2$$

The strain, therefore, is expressed by u and w only :

(1.3)
$$\varepsilon = \varepsilon(t;x,z) = \frac{\partial u}{\partial x} - z\frac{\partial^2 w}{\partial x^2} + \frac{1}{2}\left(\frac{\partial w}{\partial x}\right)^2$$

To write down the governing differential equations, we need a stress - strain relation. We assume that this is given as

(1.4)
$$\dot{\sigma} = \begin{cases} k\dot{\varepsilon} & \text{in the elastic region} \\ (1-\xi)k\dot{\varepsilon} & \text{in the plastic region,} \end{cases}$$

where $\dot{\sigma} = \partial\sigma/\partial t$, and k and ξ ($0<\xi<1$) are given, positive constants. The initial yielding is given by

$$|\sigma| = z_0,$$

where z_0 is a given positive constant. We say that the σ - ε relation is admissible when the same conditions as in § 2.2 of Chapter 2 are satisfied by $\dot{\sigma}$ and $\dot{\alpha}$ (see page 19).

The governing equations are derived from the principle of virtual work for dynamic problems. The virtual work δW due to the virtual strain $\delta\varepsilon$ under the stress σ is given by

$$\delta W = \int_{V_0} \sigma \delta\varepsilon \, dxdydz. \tag{1.5}$$

We introduce N and M by

$$N = \int_Z \sigma \, dz, \qquad M = \int_Z \sigma z \, dz. \tag{1.6}$$

Then, substituting (1.3) into (1.5) and integrating by parts, we have

$$\delta W = b \int_\Omega (N \frac{\partial \delta u}{\partial x} - M \frac{\partial^2 \delta w}{\partial x^2} + N \frac{\partial w}{\partial x} \frac{\partial \delta w}{\partial x}) \, dx$$
$$= - b \int_\Omega (\frac{\partial N}{\partial x} \delta u + \frac{\partial^2 M}{\partial x^2} \delta w + \frac{\partial}{\partial x}(N \frac{\partial w}{\partial x}) \delta w) dx.$$

We used the clamped boundary condition $\delta u = \delta w = \frac{\partial \delta w}{\partial x} = 0$ at the ends of Ω. On the other hand, the kinematic energy K is written as

$$K = \frac{1}{2} \int_{V_0} [\rho(\dot{U})^2 + \rho(\dot{W})^2] \, dxdydz,$$

where ρ is the mass density, assumed to be constant. Hence, using (1.1), the first variation of K is written as

$$\delta K = b \int_\Omega \rho [a\dot{u}\delta\dot{u} - \frac{a^3}{12} \frac{\partial^2 \dot{w}}{\partial x^2} \delta\dot{w} + a\dot{w} \, \delta\dot{w}] \, dx.$$

The principle of virtual work is expressed as

$$\int_{t_1}^{t_2} (\delta K - \delta W + \delta W_e) \, dt = 0 \qquad \text{for all } t_1, t_2 \in I,$$

where δW_e is the first variation of the external work, and $I = (0,T)$ $(T<\infty)$ is the time interval in which the problem is considered. The governing equations thus take the following form :

$$(1.7) \quad \begin{cases} \rho a\ddot{u} - \dfrac{\partial N}{\partial x} = f_x & \text{in } I\times\Omega, \\ \rho a\ddot{w} - \rho \dfrac{a^3}{12} \dfrac{\partial^2 \ddot{w}}{\partial x^2} - \dfrac{\partial^2 M}{\partial x^2} - \dfrac{\partial}{\partial x}(N \dfrac{\partial w}{\partial x}) = f_z & \text{in } I\times\Omega, \end{cases}$$

where f_x and f_z are the functions derived from the external forces. We assume that f_x and f_z are piecewise analytic with respect to t. This means that, for fixed x, these are continuous on t, and that there is a partition of I, $0 = t_0 < t_1 < \ldots , < t_I = T$, such that on each $[t_i, t_{i+1}]$ f_x (or f_z) is the restriction on this interval of some $\tilde{f}_x$ (or $\tilde{f}_z$) which is defined for $(t,x) \in D_i \times \Omega$ and analytic in D_i for any $x \in \Omega$, where D_i is a certain domain including $[t_i, t_{i+1}]$ in the complex plane. With respect to the variable x, we assume that the L^2norms of $\tilde{f}_x$ and $\tilde{f}_z$ are uniformly bounded in each D_i. The initial conditions are given by $u(0,x) = w(0,x) = 0$, $\dot{u}(0,x) = u_0$, and $\dot{w}(0,x) = w_0$, where u_0 and w_0 satisfy the clamped boundary conditions and are assumed to be sufficiently smooth. The original problem consists of (1.3)~(1.7) with the initial and boundary conditions.

10.2. Continuation of the semidiscrete solutions

To construct a solution, we start from an analysis of a finite element approximation to this problem. We divide Ω and Z at the points $\{x_p\}$ (p=0,1, 2,..., P) and at $\{z_r\}$ ($r=\pm R, \pm(R-1), \ldots, \pm 1, 0$) respectively, where $x_0 = 0$, $x_P = L$, $z_{\pm R} = \pm a/2$ and $z_0 = 0$. We assume that $x_{p+1} - x_p = h$, $z_{r+1} - z_r = \bar{h}$, and $h/\bar{h}$ is constant as h tends to 0. Let $\{x_q^0\}$ be another set of points defined as

$$x_0^0 = 0,\ x_1^0 = h/2,\ x_{q+1}^0 - x_q^0 = h \quad (q \neq 0,P),\ x_P^0 = L - h/2,\ x_{P+1}^0 = L.$$

We use P_h and Q_h to denote the sets of integers $(0,1,2,..,P)$ and $(0,1,2,..,P+1)$, respectively. Also we use P_h^0 and Q_h^0 to denote the sets obtained by excluding both the first and last numbers from P_h and Q_h. The set $[x_q^0, x_{q+1}^0] \times [z_r, z_{r+1}]$ in $\Omega \times Z$ space is called a finite element and denoted generically by e.

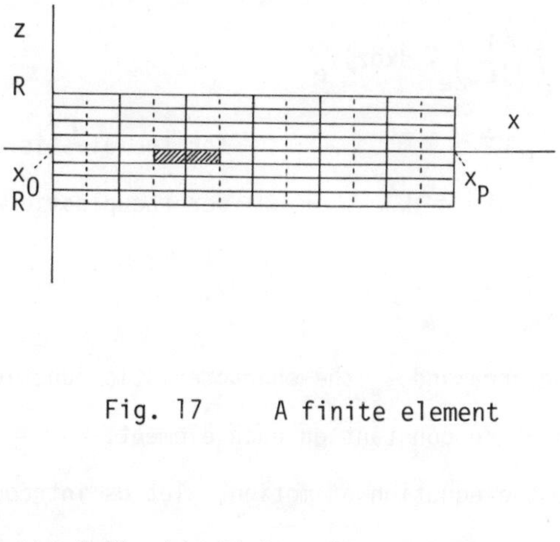

Fig. 17 A finite element

We shall use the following finite element bases :

(1). $\{\phi_p\}$ $(p \in P_h)$ is the system of continuous functions satisfying $\phi_p(x_p) = 1$, $\phi_p(x) = 0$ for $x \notin (x_{p-1}, x_{p+1})$, and linear in each $[x_i, x_{i+1}]$, where $x_{-1} = -h$, $x_{P+1} = L + h$.

(2). $\{\psi_q\}$ $(q \in Q_h)$ is the system of continuous functions satisfying $\psi_q(x_q^0) = 1$, $\psi_q(x) = 0$ for $x \notin (x_{q-1}^0, x_{q+1}^0)$, and linear in each $[x_i^0, x_{i+1}^0]$, where $x_{-1} = -h/2$, $x_{P+2} = L + h/2$.

(3). $\{\chi_p\}$ $(p \in P_h)$ is the set of the characteristic functions of the interval $[x_p - h/2, x_p + h/2]$.

Functions represented by these bases are denoted as $\hat{u}$, $\tilde{u}$ and $\bar{u}$, respectively.

The displacements u and w are approximated by $\tilde{u}$ and $\hat{w}$ in the form

$$\tilde{u} = \sum_{q \in Q_h^o} u_q(t)\psi_q , \qquad \hat{w} = \sum_{p \in P_h^o} w_p(t)\phi_p .$$

N and M are approximated by $\bar{N}$ and $\bar{M}$ constructed as follows : Let (,) and $\|\cdot\|$ be the inner product and norm in $L^2(\Omega)$.

(2.1) $$\left(\frac{\partial \hat{w}}{\partial x}, \frac{\partial \phi_p}{\partial x}\right) + (\bar{W}, \chi_p) = 0 \qquad \text{for all } p \in P_h .$$

(2.2) $$\bar{\bar{\varepsilon}} = \frac{\partial \tilde{u}}{\partial x} - z\bar{W} + \frac{1}{2}\left(\frac{\partial \hat{w}}{\partial x}\right)^2$$

(2.3) $$\bar{\varepsilon} = \sum_e \left(\frac{1}{|e|}\int_e \bar{\bar{\varepsilon}}\, dxdz\right)\chi_e$$

(2.4) $$\dot{\bar{\sigma}} = \begin{cases} k\dot{\bar{\varepsilon}} & \text{for the elastic element} \\ (1 - \xi)k\dot{\bar{\varepsilon}} & \text{for the plastic element} \end{cases}$$

(2.5) $$\bar{N} = \int_Z \bar{\sigma}\, dz, \qquad \bar{M} = \int_Z \bar{\sigma}\, z\, dz,$$

where $|e|$ denotes the area and χ_e the characteristic function of the element e. Note that $\bar{\varepsilon}$ and $\bar{\sigma}$ are constant on each element.

To approximate the equation of motion, let us introduce the operators K_i, $\bar{V}_i$ and $\bar{H}_i$ defined as

$$< K_1(u),\phi> = (\rho a\ddot{u},\phi), \quad < K_2(w),\phi> = (\rho a\ddot{w},\phi) + \frac{\rho a^3}{12}\left(\frac{\partial \ddot{w}}{\partial x},\frac{\partial \phi}{\partial x}\right)$$

$$< \bar{V}_1(u,w),\phi> = \left(\bar{N},\frac{\partial \phi}{\partial x}\right), \qquad < \bar{V}_2(u,w),\phi> = \left(\bar{N}\frac{\partial w}{\partial x},\frac{\partial \phi}{\partial x}\right) + \left(\frac{\partial \hat{M}}{\partial x},\frac{\partial \phi}{\partial x}\right)$$

$$< \bar{H}_1(u,w),\phi> = < K_1(u),\phi> + <\bar{V}_1(u,w),\phi>$$

$$< \bar{H}_2(u,w),\phi> = < K_2(w),\phi> + <\bar{V}_2(u,w),\phi>,$$

where $\hat{M} = \sum_{p \in P_h} M_p \phi_p$, if $\bar{M} = \sum_{p \in P_h} M_p \chi_p$.

The unknowns $\{u_q(t)\}$ and $\{w_p(t)\}$ are determined by solving the following system of ordinary differential equations :

$$(2.6) \quad \begin{cases} \langle \bar{H}_1(\tilde{u},\hat{w}),\psi_q \rangle = (f_x,\psi_q) & \text{for all } q \in Q_h^o \\ \langle \bar{H}_2(\tilde{u},\hat{w}),\phi_p \rangle = (f_z,\phi_p) & \text{for all } p \in P_h^o. \end{cases}$$

The initial conditions for $\tilde{u}$ and $\hat{w}$ are given by interpolating the correspond - ing functions by $\{\psi_q\}$ and $\{\phi_p\}$, respectively. The initial yielding is given by $|\bar{\sigma}| = z_0$. We call (2.1)$\sim$(2.6) the semidiscrete system.

It is still open, of course, whether the σ - ε relation of the semidiscrete system can be determined admissibly. The situation, however, is essentially the same as in the small deflection case . We shall explain briefly how the σ - ε relation is determined, especially beyond the moment at which the " yielding " or " unloading " may occur.

We start from $t = 0$. All elements are elastic until $|\bar{\sigma}| = z_0$ is satisfied for a certain element e at $t = t_0$. By the energy inequality in Lemma 2 below and by a well known fact in the theory of ordinary differential equations, the semidiscrete system has a unique analytic solution in $(0,t_0)$ for the initial conditions, if f_x and f_z are analytic in this interval. The element e may or may not yield after $t = t_0$. We want to determine the next state (elastic or plastic) of all the elements. It is obvious that the state must still be elastic for the elements satisfying $|\bar{\sigma}| < z_0$ at $t = t_0$. If the next state can be determined for all the elements, then the result must depend on the sign of $\dot{\bar{\sigma}}(t_0+0)$ or, if it should vanish, on the sign of its lowest non-zero derivative at $t = t_0+0$, since the solution is analytic, at least in a small interval $[t_0,t_0+\delta)$ $(\delta > 0)$. We will show below that this sign is determined by the data before t_0 and of the data of f_x and f_z.

First of all, we note that $\dot{\bar{\varepsilon}}$ is continuous across $t = t_0$, since $\{u_q\}$ and $\{w_p\}$ belong to the C^2-class for any choice of the next state. Hence $\dot{\bar{\sigma}}(t_0+0)$ does not vanish for the elements for which $|\bar{\sigma}| = z_0$ holds with $\dot{\bar{\sigma}}(t_0-0) \neq 0$. Therefore the next state can be determined for this kind of element. Let E be the set of all elements and E_1 be the set of elements for which $|\bar{\sigma}| = z_0$ and $\dot{\bar{\sigma}} = 0$ hold at $t = t_0+0$. We can assume that the next state is already determined for $E - E_1$, though not yet for E_1. Let us use $u^{(n)}$ to denote the n - th derivative of u with respect to t. We have the following :

(a) For the elements of E_1, the sign of $\ddot{\bar{\sigma}}(t_0+0)$ is independent of the next state of E_1.

(b) $\{u_q^{(3)}(t_0+0)\}$ ($q \in Q_h^o$) and $\{w_p^{(3)}(t_0+0)\}$ ($p \in P_h^0$) are independent of the next state of E_1.

PROOF. The sign of $\ddot{\bar{\sigma}}(t_0+0)$ is the same as that of $\ddot{\bar{\varepsilon}}(t_0+0)$ since the hardening exists (i.e., $\xi \neq 1$). Hence (a) follows from the continuity of $\ddot{\bar{\varepsilon}}$. To prove (b) we note that $\{u_q^{(3)}\}$ and $\{w_p^{(3)}\}$ are determined by the derivatives of degree less than or equal to 1 of the functions $\{w_p\}$, $\bar{N}$, $\bar{M}$, f_x and f_z. However, $\dot{\bar{N}}(t_0+0)$ and $\dot{\bar{M}}(t_0+0)$ are detemined independently of the next state of E_1, since $\dot{\bar{\sigma}}(t_0+0) = 0$ holds for the elements of E_1 and thus they have no influence on the integrations in (2.5) to determine $\dot{\bar{N}}(t_0+0)$ and $\dot{\bar{M}}(t_0+0)$. This proves (b).

By (a) we can determine the next state of the elements of E_1 for which $\ddot{\bar{\sigma}}(t_0+0)$ does not vanish. This argument can be generalized to determine the next state for all the elements. Let E_k $(k \geq 1)$ be the set of elements satisfying the following three conditions :

(1) $|\bar{\sigma}(t_0)| = z_0$ holds .

(2) $\bar{\varepsilon}^{(i)}(t_0+0) = 0$ $(1 \leq i \leq k)$ holds independently of the next state of E_k.

(3) The next states for the elements of $E - E_k$ are already determined admissibly.

Now assume that

(A) for the elements of E_k, the sign of $\bar{\sigma}^{(i+1)}(t_0+0)$ $(i \leq k)$ is independent of the next state of E_k, and

(B) $\{u_q^{(i+2)}(t_0+0)\}$ $(q \in Q_h^0)$ and $\{w_p^{(i+2)}(t_0+0)\}$ $(p \in P_h^o)$, where $i \leq k$, are independent of the next state of E_k.

Then the value $\bar{\varepsilon}^{(k+1)}(t_0+0)$ for the elements of E_k is determined independently of the next state of E_k. Moreover, if we determine, according to the sign of $\bar{\varepsilon}^{(k+1)}(t_0+0)$, the next state of the elements of E_k for which $\bar{\varepsilon}^{(k+1)}(t_0+0) \neq 0$, then the conditions (A) and (B) hold, replacing k with k+1, for all the elements of E_k which satisfy $\bar{\varepsilon}^{(k+1)}(t_0+0) = 0$ (i.e., for the elements of E_{k+1}).

PROOF. The first assertion is evident by (B), since $\bar{\varepsilon}^{(k+1)}(t_0+0)$ is determined by the derivatives of degree less than or equal to k+1 of $\{u_q\}$ and $\{w_p\}$ at $t = t_0+0$. To prove (A) for k+1, we first note that the sign of $\bar{\sigma}^{(k+2)}(t_0+0)$ is the same as that of $\bar{\varepsilon}^{(k+2)}(t_0+0)$. By the assumption (B) this is independent of the next state of E_k and especially $E_{k+1}(\subset E_k)$. This proves (A). Now $\{u_q^{(k+3)}\}$ and $\{w_p^{(k+3)}\}$ are determined by the derivatives of degree less than or equal to k+1 of the functions $\{w_p\}$, $\bar{N}$, $\bar{M}$, f_x and f_z. For the elements of E_{k+1}, it holds that $\bar{\sigma}^{(i)}(t_0+0) = 0$ $(1 \leq i \leq k+1)$, and thus $\bar{N}^{(k+1)}(t_0+0)$, etc., are independent of the next state of E_{k+1}. Also, the next state of the elements of $E - E_{k+1}$ is already determined. Hence (B) holds for k+1.

Now it is already proved that (A) and (B) hold for k = 1. Hence if E_2 is not empty, then the conditions (A) and (B) are satisfied for k = 2, and the next state is determined for the elements of $E - E_3$. If, moreover, E_3 is not empty, we can repeat this argument to get E_4, E_5,..., etc. . If E_k is empty for finite k, the next state is determined correctly for all the elements. However,there might be such elements that this repetition does not end. In that case, we determine that the next state is plastic by the same reasoning as before.

In this way, we can assign the next state of all the elements. Then the assumption on the analyticity of the solution is of course satisfied, and the solution of the initial value problem continuated at $t = t_0$ behaves as we expected, and the new $\bar{\sigma} - \bar{\varepsilon}$ relation is admissible.

For a plastic element to return to the elastic state, there must be such $t = t_1$ that $\dot{\bar{\sigma}}(t_1) = 0$ by the continuity of $\dot{\bar{\varepsilon}}$. Whether this change takes place or not is again dependent on the sign of $\ddot{\bar{\sigma}}$ or its higher derivative. The situation is, therefore, exactly the same as the first yielding, and we can again set up the next initial value problem at $t = t_1$. As is proved in Lemma 10.3 below, an energy inequality holds for the solution continuated successively by this procedure, and thus there is no bound beyond which this continuation becomes impossible.

The $\bar{\sigma} - \bar{\varepsilon}$ relation of this problem is represented by an inequality. Let $(\, , \,)_Z$ be the inner produnct and $\| \, \|_Z$ the norm of $L^2(\Omega\times Z)$. Also, by $\bar{F}$ we denote the set of functions defined on $I\times\Omega\times Z$ that are constant on each element and belong to C^1_+ with respect to t. Define $\bar{F}_{\bar{\beta}}$ to be the set of functions $\bar{\tau} =$

$\bar{\tau}(t;x,z) \in \bar{F}$ such that

$$\max_{I\times\Omega\times Z} |\bar{\tau} - \bar{\beta}| \le z_0 \qquad (\bar{\beta} \in \bar{F})$$

THEOREM 10.1. The initial value problem for the semidiscrete system is equivalent to the following problem : Let $\bar{\varepsilon}$ be defined by (2.1)~(2.3), and $\bar{N}$ and $\bar{M}$ by (2.5). Seek $\{u_q\}$, $\{w_p\} \in C^2(I)$, $\bar{\sigma} \in \bar{F}_{\bar{\alpha}}$ and $\bar{\alpha} \in \bar{F}$ which satisfy

$$(2.7) \qquad \begin{aligned} &\langle \bar{H}_1(\tilde{u},\hat{w}),\psi_q\rangle = (f_x,\psi_q) && \text{for all } q \in Q_h^o \\ &\langle \bar{H}_2(\tilde{u},\hat{w}),\phi_p\rangle = (f_z,\phi_p) && \text{for all } p \in P_h^o \end{aligned}$$

$$(2.8) \qquad (\dot{\bar{\sigma}} - k\dot{\bar{\varepsilon}}, \bar{\tau} - \bar{\sigma})_Z \ge 0 \qquad \text{for all } \bar{\tau} \in \bar{F}_{\bar{\beta}}$$

$$(2.9) \qquad \dot{\bar{\alpha}} = (1 - \frac{1}{\xi})(\dot{\bar{\sigma}} - k\dot{\bar{\varepsilon}})$$

with the same initial conditions for $\tilde{u}$, $\hat{w}$ and $\bar{\sigma}$ as in the semidiscrete system, and with $\bar{\alpha}(0) = 0$.

PROOF. We first prove that the solution for the previous problem satisfies (2.7)~(2.9), taking $\bar{\alpha}$ as the function representing the center of the yield surface (two points, in this case). Let E^e and E^p be the sets of elements in elastic and plastic states at time t. For the element $e \in E^e$, it holds that $\dot{\bar{\sigma}} - k\dot{\bar{\varepsilon}} = 0$. Since $\dot{\bar{\sigma}} - k\dot{\bar{\varepsilon}} = -\xi k\dot{\bar{\varepsilon}}$ for the plastic elements, we have

$$(\dot{\bar{\sigma}} - k\dot{\bar{\varepsilon}}, \bar{\tau} - \bar{\sigma})_Z = -\xi k \sum_{e\in E^p} (\dot{\bar{\varepsilon}}, \bar{\tau} - \bar{\sigma})_e.$$

On $e \in E^p$, it holds that $\bar{\sigma} = \bar{\alpha} + z_0$ if $\dot{\bar{\varepsilon}} > 0$ and $\bar{\sigma} = \bar{\alpha} - z_0$ if $\dot{\bar{\varepsilon}} < 0$, so $\dot{\bar{\varepsilon}}(\bar{\tau} - \bar{\sigma}) < 0$ in any case. Hence (2.8) is satisfied. To prove (2.9), we note that $\dot{\bar{\alpha}} = 0$ for the elastic elements. Since $\dot{\bar{\sigma}} = (1 - \xi)k\dot{\bar{\varepsilon}}$ and $\dot{\bar{\alpha}} = \dot{\bar{\sigma}}$ for the plastic elements, we have for the plastic elements

$$(1 - \frac{1}{\xi})(\dot{\bar{\sigma}} - k\dot{\bar{\varepsilon}}) = (1 - \xi)k\dot{\bar{\varepsilon}} = \dot{\bar{\sigma}} = \dot{\bar{\alpha}},$$

which proves (2.9).

To complete the proof it suffices to prove the uniqueness of the solution to (2.7)~(2.9). For that we prepare

LEMMA 10.2. Let $\bar{W}$ and $\bar{\varepsilon}$ be the functions determined by (2.1)~(2.3), and set $\bar{\alpha} = (1 - \frac{1}{\xi})(\bar{\sigma} - k\bar{\varepsilon})$. Then we have

$$\max_{\Omega} |\frac{\partial \hat{W}}{\partial x}| \leq C\|\bar{W}\| \tag{2.10}$$

$$\|\bar{W}\| \leq C(\|\bar{\sigma}\|_Z + \|\bar{\alpha}\|_Z) \tag{2.11}$$

$$\|\dot{\bar{W}}\| \leq C(\|\dot{\bar{\sigma}}\|_Z + \|\dot{\bar{\alpha}}\|_Z) \tag{2.12}$$

for any $t \in I$, where C is a generic constant.

PROOF. The equation (2.1) to determine $\bar{W}$ is written as

$$-\frac{1}{h}(w_{p+1} - w_p) + \frac{1}{h}(w_p - w_{p-1}) + W_p h = 0 \qquad p \in P_h$$

with evident modifications at the ends of Ω. We thus have

$$\frac{1}{h}|w_{p+1} - w_p| \leq \sum_{i=0}^{p} |W_i| h \leq C\|\bar{W}\|,$$

which proves (2.10). To prove (2.11), we note that

$$\int_Z \bar{\sigma}\, z\, dz = \int_Z (\frac{\xi}{\xi - 1}\bar{\alpha} + k\bar{\varepsilon}) z dz.$$

By (2.2)~(2.3) we have $\int_Z \bar{\varepsilon} z dz = -C\bar{W}$, where C is a positive constant depending only on a, so that

$$\bar{W} = \frac{1}{kC}\int_Z (\frac{\xi}{\xi - 1}\bar{\alpha} - \bar{\sigma}) z dz.$$

Hence (2.11) is obtained by integrating the square of both sides of this identity over Ω. (2.12) is proved in the same manner.

To simplify the expression, let us introduce the ' energy ' functionals K and E defined as

$$K(\tilde{u},\hat{w}) = \frac{1}{2}\rho a\|\dot{\tilde{u}}\|^2 + \frac{1}{2}\rho a\|\dot{\hat{w}}\|^2 + \frac{\rho a^3}{24}\|\frac{\partial \dot{\hat{w}}}{\partial x}\|^2$$

$$E(\tilde{u},\hat{w},\bar{\sigma},\bar{\alpha}) = K(\tilde{u},\hat{w}) + \frac{\xi}{2(1-\xi)k}\|\bar{\alpha}\|_Z^2 + \frac{1}{2k}\|\bar{\sigma}\|_Z^2$$

Let ($\tilde{u}_*$, $\hat{w}_*$, $\bar{\sigma}_*$, $\bar{\alpha}_*$, $\bar{\varepsilon}_*$) be another solution. The function of the form

(2.13) $$\bar{\tau} = \bar{\alpha} + z_0\bar{\theta} ,$$

where $\bar{\theta} \in \bar{F}$ and $\mathrm{Max}|\bar{\theta}| \leq 1$, belongs to $\bar{F}_{\bar{\alpha}}$. Substituting (2.13) and (2.9) into (2.8), we have

$$(\dot{\bar{\alpha}},\bar{\alpha} + z_0\bar{\theta} - \bar{\sigma})_Z \leq 0, \qquad (\dot{\bar{\alpha}}_*, \bar{\alpha}_* + z_0\bar{\theta} - \bar{\sigma}_*)_Z \leq 0$$

for all $\bar{\theta}$ as above. Hence, replacing $\bar{\theta}$ with $(\bar{\sigma}_* - \bar{\alpha}_*)/z_0$ and $(\bar{\sigma} - \bar{\alpha})/z_0$, and adding the resulting inequalities, we have

(2.14) $$(\dot{\bar{\alpha}} - \dot{\bar{\alpha}}_*,\bar{\alpha} - \bar{\alpha}_* - [\bar{\sigma} - \bar{\sigma}_*])_Z \leq 0.$$

On the other hand, by (2.9) we have

(2.15) $$\begin{aligned} &- (\dot{\bar{\alpha}} - \dot{\bar{\alpha}}_*,\bar{\sigma} - \bar{\sigma}_*)_Z \\ &= (\frac{1}{\xi} - 1)(\dot{\bar{\sigma}} - \dot{\bar{\sigma}}_*,\bar{\sigma} - \bar{\sigma}_*)_Z - (\frac{1}{\xi} - 1)k(\bar{\sigma} - \bar{\sigma}_*,\dot{\bar{\varepsilon}} - \dot{\bar{\varepsilon}}_*)_Z \end{aligned}$$

To estimate the last term, we use the system (2.7). We have the following identities :

(2.16) $$\begin{aligned} &< \bar{H}_1(\tilde{u},\hat{w}) - \bar{H}_1(\tilde{u}_*,\hat{w}_*), \dot{\tilde{u}} - \dot{\tilde{u}}_*> \\ &= <K_1(\tilde{u}) - K_1(\tilde{u}_*) + \bar{V}_1(\tilde{u},\hat{w}) - \bar{V}_1(\tilde{u}_*,\hat{w}_*), \dot{\tilde{u}} - \dot{\tilde{u}}_*> = 0. \end{aligned}$$

$$(2.17)\quad \begin{aligned} &< \bar{H}_2(\tilde{u},\hat{w}) - \bar{H}_2(\tilde{u}_*,\hat{w}_*),\ \dot{\hat{w}} - \dot{\hat{w}}_*> \\ &= < K_2(\hat{w}) - K_2(\hat{w}_*) + \bar{V}_2(\tilde{u},\hat{w}) - \bar{V}_2(\tilde{u}_*,\hat{w}_*),\ \dot{\hat{w}} - \dot{\hat{w}}_*> = 0. \end{aligned}$$

We add (2.16) and (2.17). Then we first have

$$(2.18)\quad \begin{aligned} &< K_1(\tilde{u}) - K_1(\tilde{u}_*),\ \dot{\tilde{u}} - \dot{\tilde{u}}_* > + < K_2(\hat{w}) - K_2(\hat{w}_*),\ \dot{\hat{w}} - \dot{\hat{w}}_* > \\ &= \frac{d}{dt} K(\tilde{u} - \tilde{u}_*,\ \hat{w} - \hat{w}_*). \end{aligned}$$

On the other hand, by the definition of $\bar{\varepsilon}$, we have

$$\begin{aligned} (\bar{\sigma},\bar{\varepsilon})_Z &= \sum_e (\bar{\sigma},\bar{\varepsilon})_e = \sum_e \bar{\sigma}(e)\Big(\frac{1}{|e|}\int_e \bar{\bar{\varepsilon}}\, dxdz\Big)|e|. \\ &= \sum_e (\bar{\sigma},\bar{\bar{\varepsilon}})_e = (\bar{\sigma},\bar{\bar{\varepsilon}})_Z. \end{aligned}$$

Therefore we have

$$\begin{aligned} (\bar{\sigma} - \bar{\sigma}_*, \dot{\bar{\varepsilon}} - \dot{\bar{\varepsilon}}_*)_Z &= (\bar{\sigma} - \bar{\sigma}_*, \dot{\bar{\bar{\varepsilon}}} - \dot{\bar{\bar{\varepsilon}}}_*)_Z \\ &= (\bar{\sigma} - \bar{\sigma}_*, \frac{\partial}{\partial x}[\dot{\tilde{u}} - \dot{\tilde{u}}_*] - z[\dot{\bar{W}} - \dot{\bar{W}}_*] + \frac{\partial\hat{w}}{\partial x}\frac{\partial\dot{\hat{w}}}{\partial x} - \frac{\partial\hat{w}_*}{\partial x}\frac{\partial\dot{\hat{w}}_*}{\partial x})_Z\ , \end{aligned}$$

and thus by (2.5) and (2.1) we have

$$\begin{aligned} &< \bar{V}_1(\tilde{u},\hat{w}) - \bar{V}_1(\tilde{u}_*,\hat{w}_*),\ \dot{\tilde{u}} - \dot{\tilde{u}}_* > + < \bar{V}_2(\tilde{u},\hat{w}) - \bar{V}_2(\tilde{u}_*,\hat{w}_*),\ \dot{\hat{w}} - \dot{\hat{w}}_* > \\ &= (\bar{N} - \bar{N}_*, \frac{\partial}{\partial x}[\dot{\tilde{u}} - \dot{\tilde{u}}_*]) + (\frac{\partial}{\partial x}[\hat{M} - \hat{M}_*], \frac{\partial}{\partial x}[\dot{\hat{w}} - \dot{\hat{w}}_*]) \\ &\qquad + (\bar{N}\frac{\partial\hat{w}}{\partial x} - \bar{N}_*\frac{\partial\hat{w}_*}{\partial x}, \frac{\partial}{\partial x}[\dot{\hat{w}} - \dot{\hat{w}}_*]) \\ &= (\bar{\sigma} - \bar{\sigma}_*, \dot{\bar{\varepsilon}} - \dot{\bar{\varepsilon}}_*)_Z + R, \end{aligned}$$

where

$$R = (\bar{\sigma}_* - \bar{\sigma}, \frac{\partial\dot{\hat{w}}}{\partial x}\frac{\partial}{\partial x}[\hat{w} - \hat{w}_*])_Z + (\bar{\sigma}\frac{\partial}{\partial x}[\dot{\hat{w}} - \dot{\hat{w}}_*], \frac{\partial}{\partial x}[\hat{w} - \hat{w}_*])_Z.$$

Combining this result with (2.16)~(2.18), we have

$$- (\frac{1}{\xi} - 1)k(\bar{\sigma} - \bar{\sigma}_*, \dot{\bar{\varepsilon}} - \dot{\bar{\varepsilon}}_*)_Z$$

$$= (\frac{1}{\xi} - 1)k[\frac{d}{dt} K(\tilde{u} - \tilde{u}_*, \hat{w} - \hat{w}_*) + R].$$

Therefore, by (2.4) and (2.15), we have

$$0 \geq \frac{d}{dt} E(t) + R, \tag{2.19}$$

where

$$E(t) = E(\tilde{u} - \tilde{u}_*, \hat{w} - \hat{w}_*, \bar{\sigma} - \bar{\sigma}_*, \bar{\alpha} - \bar{\alpha}_*)(t).$$

Since $\|\frac{\partial \dot{\hat{w}}}{\partial x}\|$ and $\|\bar{\sigma}\|_Z$ are uniformly bounded, as proved in Lemma 10.3 below , we have by Lemma 10.2

$$|(\bar{\sigma}_* - \bar{\sigma}, \frac{\partial \dot{\hat{w}}}{\partial x} \frac{\partial}{\partial x}[\hat{w} - \hat{w}_*])_Z| \leq C(\|\bar{\sigma} - \bar{\sigma}_*\|_Z^2 + \|\bar{W} - \bar{W}_*\|^2),$$

$$|(\bar{\sigma} \frac{\partial}{\partial x}[\dot{\hat{w}} - \dot{\hat{w}}_*], \frac{\partial}{\partial x}[\hat{w} - \hat{w}_*])_Z| \leq C(\|\frac{\partial}{\partial x}(\dot{\hat{w}} - \dot{\hat{w}}_*)\|^2 + \|\bar{W} - \bar{W}_*\|^2)$$

for a suitable constant C. Therefore, again by Lemma 10.2, $|R| \leq CE(t)$ for a certain constant C, and thus we have by (2.19)

$$\frac{d}{dt} E(t) \leq CE(t),$$

which implies that $E(t) = 0$ since $E(0) = 0$. This completes the proof.

The first energy inequality is easily obtained.

LEMMA 10.3. Let $(\tilde{u}, \hat{w}, \bar{\sigma}, \bar{\alpha})$ be the solution of the semidiscrete system. Then for any $t \in I$ it holds that

$$E(\tilde{u}, \hat{w}, \bar{\sigma}, \bar{\alpha}) \leq C,$$

where C is a constant which depends only on the given data.

PROOF. The proof is essentially the same as the uniqueness proof in Theorem 10.1. In the inequality (2.8) we can set $\bar{\tau} = \bar{\alpha}$. Hence, by (2.9)

$$0 \geq (\dot{\bar{\sigma}} - k\dot{\bar{\varepsilon}}, \bar{\sigma} - \bar{\alpha})_Z$$

$$= \frac{\xi}{1-\xi} (\dot{\bar{\alpha}}, \bar{\alpha})_Z + (\dot{\bar{\sigma}} - k\dot{\bar{\varepsilon}}, \bar{\sigma})_Z$$

$$= \frac{1}{2} \frac{\xi}{1-\xi} \frac{d}{dt} \|\bar{\alpha}\|_Z^2 + \frac{1}{2} \frac{d}{dt} \|\bar{\sigma}\|_Z^2 - k(\dot{\bar{\varepsilon}}, \bar{\sigma})_Z$$

On the other hand, we have by (2.7)

$$(f_x, \dot{\tilde{u}}) + (f_z, \dot{\hat{w}})$$

$$= < \bar{H}_1(\tilde{u}, \hat{w}), \dot{\tilde{u}} > + < \bar{H}_2(\tilde{u}, \hat{w}), \dot{\hat{w}} >$$

$$= \frac{d}{dt} K(\tilde{u}, \hat{w}) + (\dot{\bar{\varepsilon}}, \bar{\sigma})_Z,$$

which follows from the identity

$$(\bar{N}, \frac{\partial \dot{\tilde{u}}}{\partial x}) + (\bar{N}\frac{\partial \hat{w}}{\partial x}, \frac{\partial \dot{\hat{w}}}{\partial x}) + (\frac{\partial \hat{M}}{\partial x}, \frac{\partial \dot{\hat{w}}}{\partial x})$$

$$= (\bar{\sigma}, \dot{\bar{\bar{\varepsilon}}})_Z = (\bar{\sigma}, \dot{\bar{\varepsilon}})_Z .$$

Therefore we have

$$\frac{d}{dt} E(\tilde{u}, \hat{w}, \bar{\sigma}, \bar{\alpha}) \leq (f_x, \dot{\tilde{u}}) + (f_z, \dot{\hat{w}}),$$

and thus the theorem follows.

The estimates of the higher derivatives are given by

LEMMA 10.4. Let $(\tilde{u}, \hat{w}, \bar{\sigma}, \bar{\alpha})$ be the solution of the semidiscrete system. Then for any $t \in I$, it holds that

$$E'(t) \equiv E(\dot{\tilde{u}}, \dot{\hat{w}}, \dot{\bar{\sigma}}, \dot{\bar{\alpha}}) \leq C,$$

where C is a constant which depends only on the given data.

PROOF. Let $T_i = (t_{i-1}, t_i)$ be a time interval in which no change of state occurs for any element. In this interval we can differentiate the both sides of (2.7) :

$$(2.20) \qquad \frac{d}{dt} \langle \bar{H}_1(\tilde{u},\hat{w}), \psi_q \rangle = (\dot{f}_x, \psi_q)^{\cdot} \qquad \text{for any } q \in Q_h^o$$

$$(2.21) \qquad \frac{d}{dt} \langle \bar{H}_2(\tilde{u},\hat{w}), \phi_p \rangle = (\dot{f}_z, \phi_p) \qquad \text{for any } p \in P_h^o.$$

Therefore we have

$$(\dot{f}_x, \ddot{\tilde{u}}) + (\dot{f}_z, \ddot{\hat{w}})$$
$$= \frac{d}{dt} K(\dot{\tilde{u}}, \dot{\hat{w}}) + (\dot{\bar{\sigma}}, \ddot{\bar{\varepsilon}})_Z - (\dot{\bar{\sigma}}, (\frac{\partial \dot{\hat{w}}}{\partial x})^2)_Z + (\bar{\sigma}, \frac{\partial \dot{\hat{w}}}{\partial x} \frac{\partial \ddot{\hat{w}}}{\partial x})_Z.$$

Since

$$\ddot{\bar{\varepsilon}} = \frac{1}{k} \ddot{\bar{\sigma}} + \frac{\xi}{(1-\xi)k} \ddot{\bar{\alpha}}$$

in T_i , we have

$$(\dot{\bar{\sigma}}, \ddot{\bar{\varepsilon}})_Z = \frac{1}{k} (\dot{\bar{\sigma}}, \ddot{\bar{\sigma}})_Z + \frac{\xi}{(1-\xi)k} (\dot{\bar{\alpha}}, \ddot{\bar{\alpha}})_Z.$$

Also, by Lemmas 10.2 and 10.3, we have

$$|(\dot{\bar{\sigma}}, (\frac{\partial \dot{\hat{w}}}{\partial x})^2)_Z| \leq C \|\dot{\bar{\sigma}}\|_Z \|\dot{\bar{W}}\| \|\frac{\partial \dot{\hat{w}}}{\partial x}\| \leq CE'(t),$$

$$|(\bar{\sigma}, \frac{\partial \dot{\hat{w}}}{\partial x} \frac{\partial \ddot{\hat{w}}}{\partial x})_Z| \leq C \operatorname{Max} |\frac{\partial \dot{\hat{w}}}{\partial x}| \|\bar{\sigma}\|_Z \|\frac{\partial \ddot{\hat{w}}}{\partial x}\| \leq CE'(t).$$

Therefore we have in T_i

$$\frac{d}{dt} E'(t) \leq C(E'(t) + E'(t)^{\frac{1}{2}})$$

for a certain positive constant C which is independent of the interval T_i.

Let $G(t) = E'(t)^{\frac{1}{2}}$. We then have

$$G(t_i - 0) \le [G(t_{i-1} + 0) + 1] e^{C(t_i - t_{i-1})/2} - 1.$$

Now assume that the state of an element e changes from elastic to plastic at t $= t_{i-1}$. Then, by the continuity of $\dot{\bar{\varepsilon}}$, we have

$$\frac{\xi}{2(1-\xi)k} \|\dot{\bar{\alpha}}(t_{i-1} + 0)\|_e^2 + \frac{1}{2k}\|\dot{\bar{\sigma}}(t_{i-1} + 0)\|_e^2$$

$$= \frac{\xi}{2(1-\xi)k} \|\dot{\bar{\alpha}}(t_{i-1} - 0)\|_e^2 + \frac{1-\xi}{2k}\|\dot{\bar{\sigma}}(t_{i-1} - 0)\|_e^2 .$$

On the other hand, if e changes from plastic to elastic, then $\dot{\bar{\sigma}}$ and $\dot{\bar{\alpha}}$ vanish at $t = t_{i-1} + 0$. Hence, in any situation, $G(t)$ is nonincreasing at $t = t_{i-1}$, and thus we have successively

$$G(t_i - 0) \le (G(t_{i-1} - 0) + 1) e^{C(t_i - t_{i-1})/2} - 1$$

$$\le (G(t_{i-2} + 0) + 1) e^{C(t_i - t_{i-2})/2} - 1$$

$$\le (G(0) + 1) e^{Ct_i/2} - 1.$$

Note that this argument is also valid to the case where there is such $t = t^*$ that t^* is an accumulation point of $\{t_i\}$ as above. The lemma thus follows.

10.3. The fully continuous problem

Let $\mathring{W}_2^k(Q)$ $(k \ge 1)$ be the completion of the set of all C^∞ - functions vanishing identically near the boundary of Q under the norm of the usual Sobolev space $W_2^k(Q)$. Let N and M be those defined by (1.6). We use the notations H_1 and H_2 as follows :

$$\langle H_1(u,w), \phi\rangle = \langle K_1(u), \phi\rangle + (N, \frac{\partial \phi}{\partial x})$$

$$< H_2(u,w), \phi> = < K_2(w), \phi> + (N\frac{\partial w}{\partial x}, \frac{\partial \phi}{\partial x}) - (M, \frac{\partial^2 \phi}{\partial x^2}).$$

Note that the only difference between $\overline{H}_i$ and H_i is the term including M. We also define

$$E_{\sigma,\alpha} = \{ (t,x,z) \in I\times\Omega\times Z \ ; \ |\sigma - \alpha| < z_0 \}$$

$$P_{\sigma,\alpha} = \{ (t,x,z) \in I\times\Omega\times Z \ ; \ |\sigma - \alpha| = z_0 \}$$

$$F_\beta = \{\tau \in L^\infty(I;L^2(\Omega\times Z)) \ ; \ |\tau - \beta| \leq z_0 \quad \text{a.e. } I\times\Omega\times Z \}$$

for σ, α and β in $L^\infty(I;L^2(\Omega\times Z))$.

DEFINITION. (u, w, σ, α) is a weak solution of the beam problem if

$$u \in L^\infty(I; \overset{\circ}{W}{}^1_2(\Omega)), \qquad w \in L^\infty(I; \overset{\circ}{W}{}^2_2(\Omega))$$

$$\dot{u},\ \ddot{u},\ \frac{\partial \dot{u}}{\partial x},\ \dot{w},\ \ddot{w},\ \frac{\partial \dot{w}}{\partial x} \in L^\infty(I; L^2(\Omega))$$

$$\sigma,\ \dot{\sigma},\ \alpha,\ \dot{\alpha} \in L^\infty(I; L^2(\Omega\times Z)), \quad \sigma \in F_\alpha ,$$

and it holds that a.e. on I

$$< H_1(u,w),\psi> = (f_x,\psi) \qquad \text{for all } \psi \in \overset{\circ}{W}{}^1_2(\Omega)$$

$$< H_2(u,w),\phi> = (f_z,\phi) \qquad \text{for all } \phi \in \overset{\circ}{W}{}^2_2(\Omega)$$

and

$$\begin{cases} \dot{\sigma} = k\dot{\varepsilon} & \text{a.e. } E_{\sigma,\alpha} \\ \dot{\sigma} = (1 - \xi)k\dot{\varepsilon} \ \text{ and } \ (\sigma - \alpha)\dot{\sigma}/|\sigma - \alpha| \geq 0 & \text{a.e. } P_{\sigma,\alpha} \end{cases}$$

$$\dot{\alpha} = (1 - \frac{1}{\eta})(\dot{\sigma} - k\dot{\varepsilon}) \qquad \text{a.e. } I\times\Omega\times Z$$

with the auxiliary conditions

$$N = N(t;x) = \int_Z \sigma(t;x,z)dz$$

$$M = M(t;x) = \int_Z \sigma(t;x,z)\, z\, dz$$

$$\varepsilon = \varepsilon(t;x,z) = \frac{\partial u}{\partial x} - z\frac{\partial^2 w}{\partial x^2} + \frac{1}{2}\left(\frac{\partial w}{\partial x}\right)^2 .$$

The initial conditions are given by $\sigma(0) = \alpha(0) = 0$ for σ and α , and by those in the original problem for u and w.

Our final conclusion is given by

THEOREM 10.5. The weak solution exists uniquely. It is given as the limit of the finite element solution $(\tilde{u}, \hat{w}, \bar{\sigma}, \bar{\alpha})$.

PROOF. By the energy inequalities derived in the preceding section, we have as $h \to 0$

$$\left.\begin{array}{l} \dot{\tilde{u}},\ \ddot{\tilde{u}},\ \dfrac{\partial \tilde{u}}{\partial x},\ \dfrac{\partial \dot{\tilde{u}}}{\partial x} \\ \dot{\hat{w}},\ \ddot{\hat{w}},\ \dfrac{\partial \hat{w}}{\partial x},\ \dfrac{\partial \dot{\hat{w}}}{\partial x} \\ \bar{W},\ \dot{\bar{W}},\ \bar{M},\ \bar{N} \end{array}\right\} \quad \text{remain in a bounded set of } L^\infty(I;L^2(\Omega)),$$

$$\bar{\sigma},\ \dot{\bar{\sigma}},\ \bar{\alpha},\ \dot{\bar{\alpha}},\ \bar{\varepsilon},\ \dot{\bar{\varepsilon}} \quad \text{remain in a bounded set of } L^\infty(I;L^2(\Omega\times Z)).$$

Hence we can select a subsequence $(\tilde{u}, \hat{w}, \bar{W}, \bar{M}, \bar{N})$ and $(\bar{\sigma}, \bar{\alpha}, \bar{\varepsilon})$ converging weakly* to $(u, w, \frac{\partial^2 w}{\partial x^2}, M, N)$ and $(\sigma, \alpha, \varepsilon)$, respectively, which satisfy the required smoothness and equations. The situation is thus the same as in the (geometrically) linear case considered in the preceding chapters. We emphasize here that the linear space spanned by $\{\psi_q\}$ $(q \in Q_h)$ is not monotone increasing, but this does not cause any trouble in the convergence proof. Also, to prove the convergence of $\bar{\varepsilon}$ and of the first term appearing in $\bar{V}_2$, we note that $\{\frac{\partial \hat{w}}{\partial x}\}$ can be regarded as a strongly convergent sequence in $L^2(I\times\Omega)$. In fact, it is clear that a certain subsequence $\{\hat{w}\}$ converges to w strongly and that $\{\frac{\partial \hat{w}}{\partial x}\}$ converges weakly to $\frac{\partial w}{\partial x}$ in $L^2(I\times\Omega)$. On the other hand, by (2.1) we have

$$(\frac{\partial \hat{w}}{\partial x},\frac{\partial \hat{w}}{\partial x}) + (\bar{W},\bar{w}) = 0.$$

Therefore we have

$$\|\frac{\partial \hat{w}}{\partial x}\|^2_{L^2(I\times\Omega)} = - \int_I (\bar{W},\bar{w})dt = - \int_I (\bar{W},w)dt + \int_I (\bar{W},w - \bar{w})dt.$$

By the well known inequality $\|\hat{w} - \bar{w}\| \leq Ch\|\frac{\partial \hat{w}}{\partial x}\|$, the second term on the right side tends to zero as $h \to 0$, and so the first one converges to

$$- \int_I (\frac{\partial^2 w}{\partial x^2},w)dt = \|\frac{\partial w}{\partial x}\|^2_{L^2(I\times\Omega)}$$

This implies that $\{\frac{\partial \hat{w}}{\partial x}\}$ converges strongly to $\frac{\partial w}{\partial x}$ in $L^\infty(I\times\Omega)$ as $h \to 0$. The convergence of the term including $\hat{M}$ in $\bar{V}_2$ is proved by noticing that if $\hat{\phi}$ is the interpolate of ϕ by $\{\hat{\phi}_p\}$, where ϕ is sufficiently smooth in Ω and vanishes at the ends of Ω with its first derivative, then it holds that

$$(\frac{\partial \hat{M}}{\partial x},\frac{\partial \hat{\phi}}{\partial x}) = - (\bar{M}, \frac{\partial^2 \phi}{\partial x^2}) + \gamma_h,$$

where $\gamma_h = O(h\|\bar{M}\|)$. The uniqueness of the solution is proved the same way as Theorem 10.1. Hence the theorem follows.

APPENDIX (A)

AN ELEMENTARY PROOF OF KORN'S INEQUALITY

To prove the existence of the semidiscrete or the fully continuous solutions, we required the positivity of the strain energy :

(1) $$\|\nabla u\| \leq C\|\varepsilon(u)\| \qquad \text{for all } u \in W_2^1(\Omega) \ ; \ u|_{\Gamma_u} = 0,$$

where $\|\cdot\|$ is the $L^2(\Omega)$-norm of a vector function. Such constant C always exists for the class of piecewise linear functions on a fixed finite element partition of Ω, but it might depend on the finite element parameter h. Korn's inequality assures that this constant depends only on the domain Ω. In fact, if u vanishes on the whole boundary, then it is easy to show that

(2) $$\|\nabla u\| \leq \sqrt{2}\,\|\varepsilon(u)\|.$$

To see this, let $u = (u_1, u_2)$ be a $C^\infty(\Omega)$-function vanishing identically near the boundary of Ω. By Green's theorem we have

$$\int_\Omega u_{1,2}u_{2,1}dx = -\int_\Omega u_{1,21}u_2\,dx = \int_\Omega u_{1,1}\,u_{2,2}dx,$$

so that

(3) $$2\left|\int_\Omega u_{1,2}\,u_{2,1}dx\right| \leq \int_\Omega |u_{1,2}u_{2,1}|\,dx + \int_\Omega |u_{1,1}u_{2,2}|dx \leq \frac{1}{2}\|\nabla u\|^2.$$

Now since

$$\|\varepsilon(u)\|^2 = \int_\Omega (u_{1,1}^2 + u_{2,2}^2 + (u_{1,2} + u_{2,1})^2)dx$$

$$= \|\nabla u\|^2 + 2\int_\Omega u_{1,2}u_{2,1}dx,$$

(2) is obtained by substituting (3) into this identity. Hence (2) holds for all $W_2^1(\Omega)$ – functions vanishing on the boundary of Ω (Korn's first inequality).

Korn's second inequality, i.e., an inequality of type (1), follows from the inequality

$$(4) \qquad \|\nabla u\| \le C(\|\varepsilon(u)\| + \|u\|) \qquad \text{for all } u \in W_2^1(\Omega).$$

There are several proofs of this inequality. For example, [4] and [7]. Here we shall give an elementary proof following the idea of Nitsche [18]. We note that $\|\varepsilon(u)\|$ is not invariant under rotation of the co-ordinate axis. For example, $\|\varepsilon(u)\|_x = 0$ if $u = (ax_2, -ax_1)$ $(a \neq 0)$, but $\|\varepsilon(u)\|_y \neq 0$ for any nontrivial orthogonal transformation $x = Py$. Hence we treat the boundary of Ω without local co-ordinate transformation.

We assume that the boundary Γ of the domain Ω has the piecewise C^2 - property defined in Chapter 9. Then for any boundary point p there exists a circle $S_r(p)$ of radius r and centered at p, which satisfies the following conditions : Set

$$C_p = \Gamma \cap (\text{ inside of } S_r(p)).$$

Then C_p intersects with $S_r(p)$ at two different points and separates the inside of $S_r(p)$ into the inside and outside of Ω. Moreover, taking a smaller r, if necessary, C_p satisfies one of the following four conditions (not necessarily exclusive).

1. C_p is of A or B type.
2. C_p is represented by a single-valued function and consists of two curves

of type A or two curves of type B.

3. C_p cannot be represented by a single-valued function but consists of two curves of type A or type B.

4. C_p consists of two curves, one of type A and the other of type B.

Some typical cases of C_p are illustrated below.

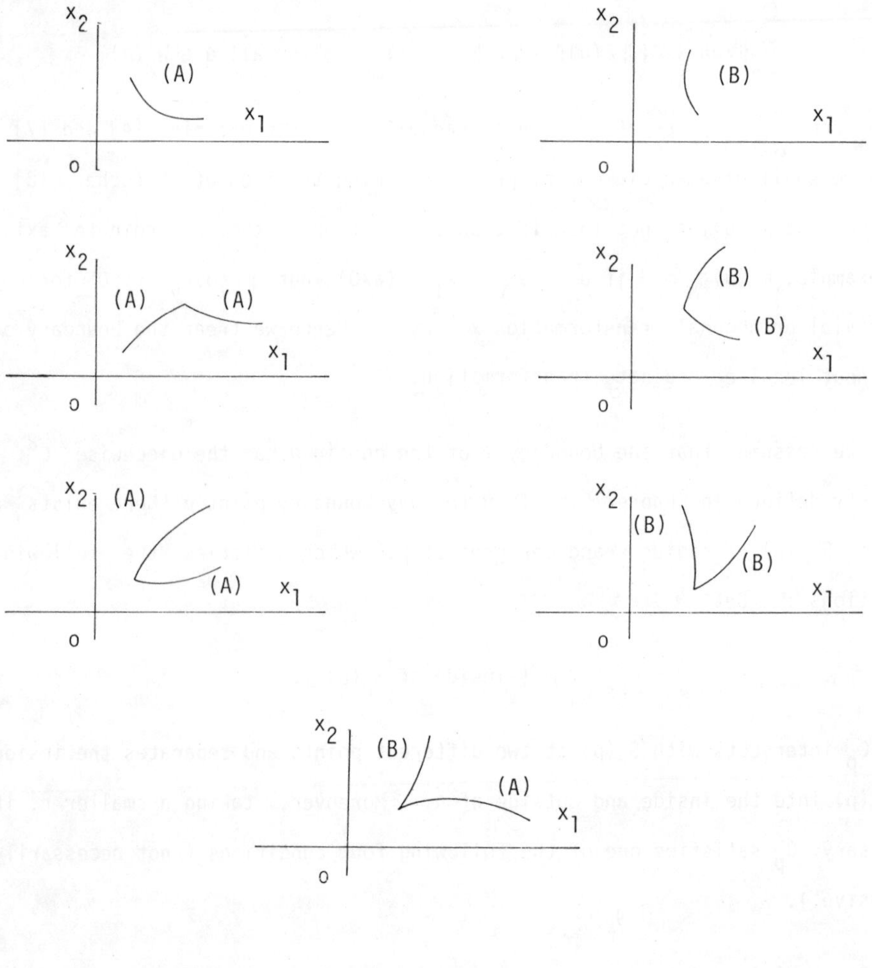

Fig. 18 Types of a portion of the boundary

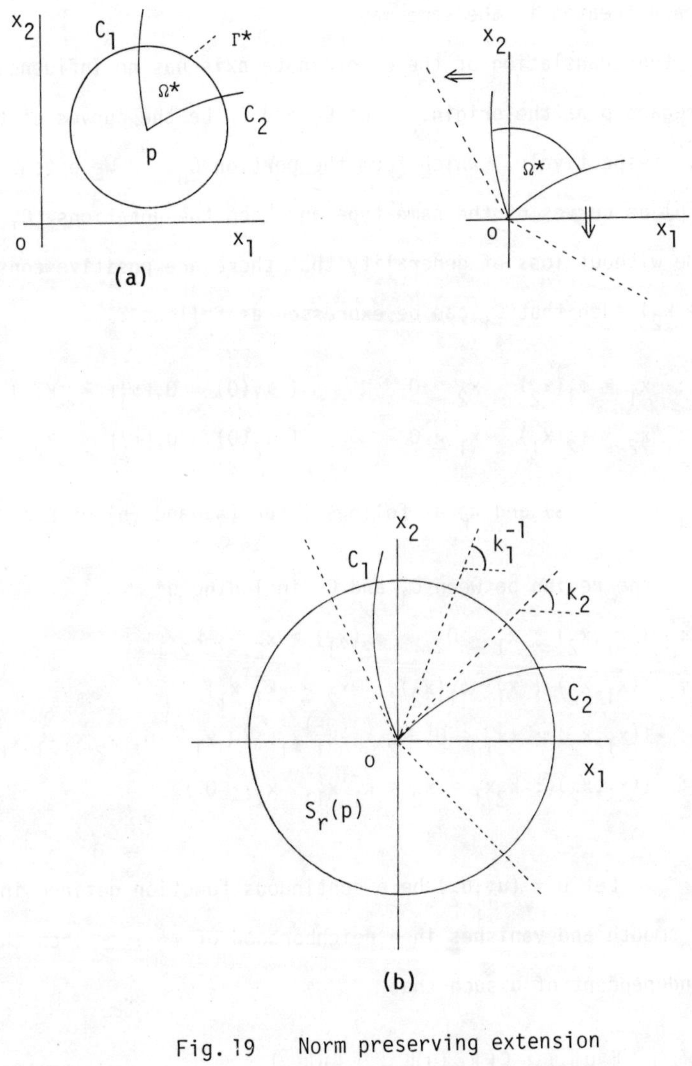

Fig. 19 Norm preserving extension

Let $S_r(p)$ be the circle defined above for a boundary point p, and set Ω^* $= \Omega \cap$ (inside of $S_r(p)$), $\Gamma^* = \Omega \cap S_r(p)$. We first prove inequality (4) for the functions defined in Ω^* and which vanish identically near Γ^*. The proof is done only for the case satisfying condition 4 above, but as will be seen, the other cases are treated in the same way.

Since the translation of the co-ordinate axis has no influence on the result, we regard p as the origin. Let C_1 and C_2 be the curves of type B and of type A, respectively, which form the portion C_p. We extend C_1 and C_2 outside $S_r(p)$ as curves of the same type and keep the notations C_1 and C_2. We can assume without loss of generality that there are positive constants k_1 and k_2 ($k_1^{-1} > k_2$) such that C_i can be expressed as follows :

$$C_1 : \ x_1 = \phi_1(x_2) \quad x_2 \geq 0 \qquad (\ \phi_1(0) = 0, |\phi_1'| \leq k_1\ )$$
$$C_2 : \ x_2 = \phi_2(x_1) \quad x_1 \geq 0 \qquad (\ \phi_2(0) = 0, |\phi_2'| \leq k_2\ ).$$

We define $\Omega_i (i = 0 \sim 3)$ and Ω_K as follows (see (a) and (b) of Fig. 19):

Ω_0 : the region between C_1 and C_2 including Ω^*

Ω_1 : $\{(x_1,x_2)\ ;\ x_1 \geq 0 \quad \phi_2(x_1) > x_2 \geq -k_2x_1\}$

Ω_2 : $\{(x_1,x_2)\ ;\ x_1 < \phi_1(x_2), \quad x_2 \geq -k_1^{-1}x_1\}$

Ω_3 : $\{(x_1,x_2)\ ;\ \{x_1 < 0,\ x_2 < -k_1^{-1}x_1\} \cup \{x_1 > 0,\ x_2 < -k_2x_1\}$

Ω_K : $\{(x_1,x_2)\ ;\ k_2x_1 < x_2 < k_1^{-1}x_1,\ \ x_1 \geq 0\ \}$

THEOREM A1. Let $u = (u_1,u_2)$ be a continuous function defined in Ω^* which is piecewise smooth and vanishes in a neighborhood of Γ^*. Then there is a constant C independent of u such that

$$\text{(5)} \qquad \|\nabla u\|_{\Omega^*} \leq C(\ \|\varepsilon(u)\|_{\Omega^*} + \|u\|_{\Omega^*}).$$

PROOF. The proof consists of three steps.

(i). There exists an extension of u as a piecewise smooth function from Ω^* to $\Omega_0 \cup \Omega_1$, and it holds that

$$\text{(6)} \qquad \|\varepsilon(u)\|_{\Omega_0 \cup \Omega_1} \leq C(\|\varepsilon(u)\|_{\Omega^*} + \|u\|_{\Omega^*}), \qquad \|u\|_{\Omega_0 \cup \Omega_1} \leq C\|u\|_{\Omega^*}.$$

To prove this, we regard u as a function defined on Ω_0 by setting u=0 in $\Omega_0 - \Omega^*$. By u_i^λ ($\lambda > 0$) we denote the function

$$u_i^\lambda(x) = u_i(x_1, \phi_2(x_1) + \lambda[\phi_2(x_1) - x_2]) \qquad x \in \Omega_1.$$

Owing to the domain Ω_K, the point

$$(x_1, \phi_2(x_1) + \lambda[\phi_2(x_1) - x_2])$$

is included in Ω_0, provided λ is sufficiently small (in fact, $\lambda < (k_1^{-1} - k_2)/2k_2$ is sufficient), and hence u_i^λ is well defined. We shall seek an extension of u to Ω_1 in the form

$$u_1(x) = pu_1^\lambda + qu_1^\mu + \rho(x_1)u_2^\lambda + \sigma(x_1)u_2^\mu$$

$$u_2(x) = ru_2^\lambda + su_2^\mu .$$

The continuity and piecewise smoothness are ensured by the conditions

$$\text{(7)} \qquad p + q = r + s = 1, \quad \rho(x_1) + \sigma(x_1) = 0.$$

We try to express $\varepsilon_{ij}(u)$ by a linear combination of $\varepsilon_{ij}^\lambda(u)$ and u_i^λ, etc. Since

$$\begin{aligned} \varepsilon_{11}(u) = {} & pu_{1,1}^\lambda + pu_{1,2}^\lambda(1+\lambda)\phi_2' + qu_{1,1}^\mu + qu_{1,2}^\mu(1+\mu)\phi_2' \\ & + \rho u_{2,1}^\lambda + \rho u_{2,2}^\lambda(1+\lambda)\phi_2' + \sigma u_{2,1}^\mu + \sigma u_{2,2}^\mu(1+\mu)\phi_2' \\ & + \rho' u_2^\lambda + \sigma' u_2^\mu , \end{aligned}$$

the conditions

$$(8) \qquad p(1+\lambda)\phi_2' = \rho \ , \qquad q(1+\mu)\phi_2' = \sigma$$

are sufficient to express

$$(9) \qquad \varepsilon_{11}(u) = p\varepsilon_{11}^{\lambda}(u) + q\varepsilon_{11}^{\mu}(u) + \rho(1+\lambda)\phi_2'\varepsilon_{22}^{\lambda}(u) + \sigma(1+\mu)\phi_2'\varepsilon_{22}^{\mu}(u)$$

$$+ \rho\varepsilon_{12}^{\lambda}(u) + \sigma\varepsilon_{12}^{\mu}(u) + \rho' u_2^{\lambda} + \sigma' u_2^{\mu} \ .$$

Also, since

$$\varepsilon_{22}(u) = -r\lambda u_{2,2}^{\lambda} - s\mu u_{2,2}^{\mu},$$

we have

$$(10) \qquad \varepsilon_{22}(u) = -r\lambda\varepsilon_{22}^{\lambda}(u) - s\mu\varepsilon_{22}^{\mu}(u).$$

Finally, the condition

$$(11) \qquad -p\lambda = r, \qquad -q\mu = s$$

allow us to write

$$(12) \qquad \varepsilon_{12}(u) = r\varepsilon_{12}^{\lambda}(u) + s\varepsilon_{12}^{\mu}(u)$$

$$+ [\ r(1+\lambda)\phi_2' - \rho\lambda]\ \varepsilon_{22}^{\lambda}(u)$$

$$+ [\ s(1+\mu)\phi_2' - \sigma\mu]\ \varepsilon_{22}^{\mu}(u) \ .$$

If conditions (7),(8) and (11) are satisfied, then (6) follows from (9), (10) and (12). We choose the parameters as follows : First λ is chosen so small that u^{λ} is well defined. Then it is sufficient to set

$$p = -\frac{2}{\lambda} \ , \quad q = 1 + \frac{2}{\lambda} \ , \quad r = 2 \ , \quad s = -1 \ , \quad \mu = \frac{\lambda}{2+\lambda}$$

$$\rho(x_1) = p(1+\lambda)\phi_2'(x_1), \qquad \sigma(x_1) = q(1+\mu)\phi_2'(x_1).$$

(2). There exists an extension of u from Ω^* to $\Omega_0 \cup \Omega_2$ as a piecewise smooth function, and it holds that

$$(13) \qquad \|\varepsilon(u)\|_{\Omega_0\cup\Omega_2} \le C(\|\varepsilon(u)\|_{\Omega^*} + \|u\|_{\Omega^*}), \qquad \|u\|_{\Omega_0\cup\Omega_2} \le C\|u\|_{\Omega^*}.$$

The proof is now easy. We replace $\phi_2(x_1)$ by $\phi_1(x_2)$ in the proof of (1) and seek an extension of the form

$$u_1(x) = pu_1^\lambda + qu_1^\mu$$

$$u_2(x) = ru_2^\lambda + su_2^\mu + \rho(x_2)u_1^\lambda + \sigma(x_2)u_1^\mu,$$

where

$$u_i^\lambda(x) = u_i(\phi_1(x_2) + \lambda(\phi_1(x_2) - x_1), x_2).$$

Then the conditions

$$\lambda, \mu > 0$$
$$p + q = r + s = 1, \ \rho + \sigma = 0, \ -r\lambda = p, \ -s\mu = q$$
$$\rho = r(1+\lambda)\phi_1', \quad \sigma = s(1+\mu)\phi_1'$$

are sufficient to get (13).

Set $D = \Omega_0 \cup \Omega_1 \cup \Omega_2$. By (1) and (2) we have an extension of u from Ω^* to D which satisfies

$$\|\varepsilon(u)\|_D \le C\{\|\varepsilon(u)\|_{\Omega^*} + \|u\|_{\Omega^*}\}, \qquad \|u\|_D \le C\|u\|_{\Omega^*}.$$

(3). Let R_+^2 and R_-^2 be the upper and lower half-plane, and set

$$\Omega^+ = D \cap R^2_+ , \qquad \Omega^- = D \cap R^2_- .$$

We extend $u = u|_{\Omega^+}$, first from Ω^+ to R^2_+, and then from R^2_+ to the whole plane by the method as above. Then clearly we have

$$\|\varepsilon(u)\|_{R^2_+} \le C\{\|\varepsilon(u)\|_{\Omega^+} + \|u\|_{\Omega^+}\} \qquad \|u\|_{R^2_+} \le C\,\|u\|_{\Omega^+}$$

$$\|\varepsilon(u)\|_{R^2} \le C\{\|\varepsilon(u)\|_{R^2_+} + \|u\|_{R^2_+}\} \qquad \|u\|_{R^2} \le C\,\|u\|_{R^2_+}$$

and so

$$\begin{aligned}\|\varepsilon(u)\|_{R^2} &\le C\{\|\varepsilon(u)\|_{\Omega^+} + \|u\|_{\Omega^+}\} \\ &\le C\{\|\varepsilon(u)\|_{\Omega^*} + \|u\|_{\Omega^*}\}\end{aligned}$$

Since u has a compact support, we have by Korn's first inequality

$$\begin{aligned}\|\nabla u\|_{\Omega^+} \le \|\nabla u\|_{R^2} &\le \sqrt{2}\,\|\varepsilon(u)\|_{R^2} \\ &\le C\{\|\varepsilon(u)\|_{\Omega^*} + \|u\|_{\Omega^*}\}\end{aligned}$$

The same estimate holds for $u = u|_{\Omega^-}$. Therefore, combining the two estimates we have inequality (5).

THEOREM A2. Assume that the boundary Γ has the piecewise C^2 - property. Then there is a constant C such that

(14) $$\|\nabla u\|_{\Omega} \le C(\|\varepsilon(u)\|_{\Omega} + \|u\|_{\Omega}) \qquad \text{for all } u \in W^1_2(\Omega).$$

PROOF. It suffices to show this for piecewise smooth functions on Ω. Take $p \in \Gamma$ and let $S_r(p)$ be the circle defined above. Since the boundary is compact, it is covered by the union of the insides of $S_{r_1}(p_1), S_{r_2}(p_2), \ldots, S_{r_n}(p_n)$, for example. Let

$$1 = \sum_{i=1}^{n} \phi_i$$

be a partition of unity with respect to these $\{S_{r_i}(p_i)\}$, and set

$$u_{(i)} = \phi_i u , \qquad u_{(0)} = (1 - \sum_{i=1}^{n} \phi_i)u .$$

Korn's first inequality is then applicable to $u_{(0)}$ since its support is included in Ω. Hence (14) holds for $u_{(0)}$. On the other hand, since $u_{(i)}$ vanishes in a neighborhood of $\Gamma_i^* = \Omega \cap S_{r_i}(p_i)$, inequality (5) holds for $u_{(i)}$ and for $\Omega_i^* = \Omega \cap$(inside of $S_{r_i}(p_i)$). Hence the theorem holds well.

THEOREM A3 (Korn's second inequality). Let Γ satisfy the condition of the preceding theorem. If measure$(\Gamma_u) \neq 0$, then the following inequality holds.

$$\|\nabla u\|_\Omega \leq C\|\varepsilon(u)\|_\Omega \qquad \text{for all } u \in W_2^1(\Omega) \ ; \ u|_{\Gamma_u} = 0.$$

PROOF. Assume this is not true. Then there is a sequence u_n such that

$$\|\nabla u_n\|_\Omega > n\|\varepsilon(u_n)\|_\Omega, \qquad u_n \in W_2^1(\Omega) \ ; \ u_n|_{\Gamma_u} = 0.$$

Set $v_n = u_n/\|\nabla u_n\|_\Omega$. Then v_n satisfies

$$\|\nabla v_n\|_\Omega = 1, \quad \frac{1}{n} > \|\varepsilon(v_n)\|_\Omega \qquad \text{for all } n \geq 1.$$

Hence, there is a subsequence $\{v_n\}$ and $v \in W_2^1(\Omega)$ such that

$$v_n \to v \qquad \text{weakly in } W_2^1(\Omega), \text{ strongly in } L^2(\Omega)$$

$$\|\varepsilon(v_n)\|_\Omega \to 0.$$

Since $\|\varepsilon(v)\| \leq \underline{\lim} \|\varepsilon(v_n)\| = 0$, we have $v = 0$ by the boundary condition. On the other hand, by Theorem A2 we have

$$1 \leq C\|v\|_{\Omega}$$

which is a contradiction. This completes the proof.

REMARK. In the above proof, we used the inequality

$$\|u\|_{\Omega} \leq C\|\nabla u\|_{\Omega} \qquad \text{for all } u \in W_2^1(\Omega)\ ;\ u|_{\Gamma_u} = 0.$$

For the proof of this inequality, see [3], for example.

APPENDIX (B)

JOHNSON'S IMPLICIT METHOD

In the preceding chapters we concentrated on the analysis of explicit integration schemes, in which the hardening parameters are not regarded as the unknown. But in a method proposed by Johnson [10], these parameters are the unknowns, and a weak form of the problem is approximated directly. In this appendix we examine Johnson's method, taking the problem (5.1)∼(5.4) of Chapter 9 as an example.

The solution (u,σ,α) of the above problem satisfies a.e. I

$$\dot{\varepsilon} - C\dot{\sigma} = 0, \quad \dot{\alpha} = 0$$

or

$$\dot{\varepsilon} - C\dot{\sigma} = \frac{1}{\eta}\,\partial f \partial f^{*}\dot{\sigma}, \quad \dot{\alpha} = (\sigma - \alpha)\frac{\partial f^{*}\dot{\sigma}}{f}, \quad \partial f^{*}\dot{\sigma} \geq 0$$

almost everywhere on Ω. In the later case, it holds that

$$\begin{pmatrix} \dot{\varepsilon} - C\dot{\sigma} \\ \\ -\frac{1}{\eta}S\dot{\alpha} \end{pmatrix} = \begin{pmatrix} \frac{1}{\eta}\,\partial f\ \partial f^{*}\dot{\sigma} \\ \\ -\frac{1}{\eta}\partial f\ \partial f^{*}\dot{\sigma} \end{pmatrix} = \frac{1}{\eta}\begin{pmatrix} \partial f \\ \\ -\partial f \end{pmatrix}\partial f^{*}\dot{\sigma} \qquad (\partial f^{*}\dot{\sigma} \geq 0).$$

Therefore the solution satisfies the following relations a.e. I :

(1) $$(\dot{\varepsilon} - C\dot{\sigma}, \tau - \sigma) - \frac{1}{\eta}(S\dot{\alpha}, \zeta - \alpha) \leq 0$$

$$\text{for all } (\tau,\zeta) \in L^{\infty}(\Omega);\ f(\tau - \zeta) \leq z_0 \quad \text{a.e. } \Omega$$

(2) $$\sum_{j=1}^{2} (\sigma_{ij}, \phi_{,j}) = (b_i, \phi) \qquad \text{for all } \phi \in D_2^1(\Omega,\Gamma_u).$$

In what follows, the system (2) is written simply as

$$(\sigma,\varepsilon(\phi)) = (b,\phi).$$

We use the following notations to denote the spaces of functions. Let Ω be a polygonal domain for simplicity. We fix a triangular partition of Ω.

W_h : the set of all piecewise linear finite element functions vanishing on Γ_u.

W^0 : the set of piecewise constant finite element functions.

In Johnson's method, the equations to compute $(u^{n+1}, \sigma^{n+1}, \alpha^{n+1})$ from (u^n,σ^n,α^n) are

$$(3) \qquad (D_t\varepsilon^n - CD_t\sigma^n,\tau - \sigma^{n+1}) - \frac{1}{\eta}(SD_t\alpha^n,\zeta - \alpha^{n+1}) \leq 0$$

$$\text{for all } (\tau,\zeta) \in W^0 \ : f(\tau - \zeta) \leq z_0 ,$$

$$(4) \qquad (\sigma^{n+1},\varepsilon(w)) = (b^{n+1},w) \qquad \text{for all } w \in W_h,$$

with $f(\sigma^{n+1} - \alpha^{n+1}) \leq z_0$, where $(u^k,\sigma^k,\alpha^k) \in W_h \times W^0 \times W^0$ and $\varepsilon^k = \varepsilon(u^k)$.

The solvability and error estimate of this method are given by

THEOREM B1 (i). Problem (3)∼(4) has a unique solution.

(ii). Let (u,σ,α) be the exact solution. Then we have the error estimate

$$(5) \qquad \|\varepsilon(n\Delta t) - \varepsilon^n\| + \|\sigma(n\Delta t) - \sigma^n\| + \|\alpha(n\Delta t) - \alpha^n\| \leq C(\ \delta(h) + \sqrt{\Delta t}\ ),$$

where $\|\cdot\|$ is the $L^2(\Omega)$ norm and

$$\delta(h) = \inf_{v \in L^2(I;D^1(\Omega,\Gamma_u))} \|\varepsilon(u) - \varepsilon(v)\|_{L^2(I\times\Omega)}$$

PROOF. We apply Uzawa's iteration to approximate the solution of the system

(3)~(4) and we prove that this iteration converges. In other words, we prove the solvability of (3)~(4) by analysing a numerical method to solve (3)~(4) in actuality. To avoid complexity of expression, we use the inner product

$$(\hat{\sigma},\hat{\tau}) = (C\sigma,\tau) + \frac{1}{\eta}(S\alpha,\zeta) \qquad \text{for } \hat{\sigma} = (\sigma,\alpha) \text{ and } \hat{\tau} = (\tau,\zeta)$$

and set $v_j^k = (u_j^k - u^n)/\Delta t$. Then Uzawa's algorithm applied to (3)~(4) is as follows : Set $u_0^{n+1} = 0$ and determine the subsequent functions by

$$(6) \qquad \frac{1}{\Delta t}(\hat{\sigma}_j^{n+1} - \hat{\sigma}^n, \hat{\tau} - \hat{\sigma}_j^{n+1}) - (\varepsilon(v_{j-1}^{n+1}), \tau - \sigma_j^{n+1}) \geq 0$$

$$\text{for all } \hat{\tau} \in W^0 \;;\; f(\tau - \zeta) \leq z_0 \,,$$

$$(7) \qquad (\varepsilon(v_j^{n+1}),\varepsilon(w)) = (\varepsilon(v_{j-1}^{n+1}),\varepsilon(w)) + \rho\,((b^{n+1},w) - (\sigma_j^{n+1},\varepsilon(w)))$$

$$\text{for all } w \in W_h \,,$$

with $f(\sigma_j^{n+1} - \alpha_j^{n+1}) \leq z_0$ $(j = 1,2,..,)$. ρ is a parameter determined later. This iteration is well defined. In other words, $\hat{\sigma}_j^{n+1} = (\sigma_j^{n+1},\alpha_j^{n+1})$ is uniquely determined by (6). This is proved by the facts that for each element e the set

$$K = \{\hat{\sigma} \in W^0(e) \;;\; f(\sigma - \alpha) \leq z_0\}$$

is convex and closed in R^6 , and that the functional

$$(8) \qquad J(\hat{\sigma}) = \frac{1}{\Delta t}(\hat{\sigma},\hat{\sigma})_e - \frac{2}{\Delta t}(\hat{\sigma},\hat{\sigma}^n)_e - 2(\varepsilon(v_{j-1}^{n+1}),\sigma)_e$$

is also convex on K. As is well known, the minimizing problem of $J(\hat{\sigma})$ on K is equivalent to seeking a solution of problem(6) in K (strictly speaking, problem(6) considered only on element e), and $J(\hat{\sigma})$ has a unique minimizing point in K (see, for example,[5]).

We next show that the sequence $(u_j^{n+1},\sigma_j^{n+1},\alpha_j^{n+1})$ is convergent in $L^2(\Omega)$.

Since the finite element partition is fixed, this also implies the uniform convergence on Ω. First, we have by (7)

$$(9) \qquad (\varepsilon(v_j^{n+1}) - \varepsilon(v_{j-1}^{n}), \varepsilon(w)) = (\varepsilon(v_{j-1}^{n+1}) - \varepsilon(v_{j-2}^{n+1}), \varepsilon(w)) - \rho(\sigma_j^{n+1} - \sigma_{j-1}^{n+1}, \varepsilon(w))$$

for all $w \in W_h$. By setting $\varepsilon(w) = \varepsilon(v_j^{n+1}) - \varepsilon(v_{j-1}^{n+1})$ in (9) we have

$$(10) \qquad \| \varepsilon(v_j^{n+1}) - \varepsilon(v_{j-1}^{n+1})\|^2 = (\varepsilon(v_{j-1}^{n+1}) - \varepsilon(v_{j-2}^{n+1}), \varepsilon(v_j^{n+1}) - \varepsilon(v_{j-1}^{n+1})) - \rho(\sigma_j^{n+1} - \sigma_{j-1}^{n+1}, \varepsilon(v_j^{n+1}) - \varepsilon(v_{j-1}^{n+1})).$$

On the other hand, first set $\hat{\tau} = \hat{\sigma}_{j-1}^{n+1}$ in (6) and then $\hat{\tau} = \hat{\sigma}_j^{n+1}$ after replacing j by j-1, and add the resulting inequalities to get

$$\frac{1}{\Delta t}(\hat{\sigma}_{j-1}^{n+1} - \hat{\sigma}_j^{n+1}, \hat{\sigma}_j^{n+1} - \hat{\sigma}_{j-1}^{n+1}) - (\varepsilon(v_{j-2}^{n+1}) - \varepsilon(v_{j-1}^{n+1}), \sigma_j^{n+1} - \sigma_{j-1}^{n+1}) \geq 0$$

or equivalently

$$(11) \qquad \frac{1}{\Delta t}\| \hat{\sigma}_j^{n+1} - \hat{\sigma}_{j-1}^{n+1}\|^2 \leq (\sigma_j^{n+1} - \sigma_{j-1}^{n+1}, \varepsilon(v_j^{n+1}) - \varepsilon(v_{j-1}^{n+1})) + (\sigma_j^{n+1} - \sigma_{j-1}^{n+1}, \varepsilon(v_{j-1}^{n+1}) - \varepsilon(v_{j-2}^{n+1}) - [\varepsilon(v_j^{n+1}) - \varepsilon(v_{j-1}^{n+1})]).$$

Multiplying (11) by ρ and adding the resulting inequality to (10), we have

$$(12) \qquad \frac{1}{2}\| \varepsilon(v_j^{n+1}) - \varepsilon(v_{j-1}^{n+1})\|^2 + \frac{\rho}{\Delta t}\| \hat{\sigma}_j^{n+1} - \hat{\sigma}_{j-1}^{n+1}\|^2 \leq \frac{1}{2}\| \varepsilon(v_{j-1}^{n+1}) - \varepsilon(v_{j-2}^{n+1})\|^2 + \frac{\rho^2}{2}\| \sigma_j^{n+1} - \sigma_{j-1}^{n+1}\|^2.$$

Now since the matrix $C = D^{-1}$ is positive definite, there is a positive constant μ such that

$$(\hat{\sigma}, \hat{\sigma}) \geq \mu \ \| \sigma \|^2_{L^2(\Omega)}$$

Hence, we have by (12)

$$(13)\qquad \frac{1}{2}\| \varepsilon(v_j^{n+1}) - \varepsilon(v_{j-1}^{n+1})\|^2 + (\frac{\rho}{\Delta t} - \frac{\rho^2}{2\mu})\| \hat{\sigma}_j^{n+1} - \hat{\sigma}_{j-1}^{n+1}\|^2$$

$$\leq \frac{1}{2}\| \varepsilon(v_{j-1}^{n+1}) - \varepsilon(v_{j-2}^{n+1})\|^2.$$

Adding (13) with respect to j we see that if the condition

$$\rho < \frac{2\mu}{\Delta t}$$

is satisfied then $\{\hat{\sigma}_j^{n+1}\}$ $(j=0,1,2,..)$ is convergent in $L^2(\Omega)$. Thus there is a $\hat{\sigma}^{n+1} \in W^0$ such that $\hat{\sigma}_j^{n+1}$ converges to it uniformly. In this case, v_j^{n+1} also converges uniformly. To see this, we note that by (6) there is a constant $\lambda_j \geq 0$ such that

$$\begin{pmatrix} \varepsilon(v_{j-1}^{n+1}) - C(\sigma_j^{n+1} - \sigma^n)/\Delta t \\ \\ -\frac{1}{\eta} S(\alpha_j^{n+1} - \alpha^n)/\Delta t \end{pmatrix} = \lambda_j \begin{pmatrix} \partial f_{n+1} \\ \\ - \partial f_{n+1} \end{pmatrix}$$

That is,

$$(14)\qquad \frac{1}{\eta} S \frac{\alpha_j^{n+1} - \alpha^n}{\Delta t} = \varepsilon(v_{j-1}^{n+1}) - C\frac{\sigma_j^{n+1} - \sigma^n}{\Delta t}$$

Since $(\sigma_j^{n+1}, \alpha_j^{n+1})$ converges, $\varepsilon(v_j^{n+1})$ and hence v_j^{n+1} and u_j^{n+1} converge, e.g., to v^{n+1} and u^{n+1}, respectively. Then $(u^{n+1}, \hat{\sigma}^{n+1})$ satisfies (3)~(4).
The uniqueness of the solution and the error estimate can be shown by the method used frequently in the preceding chapters.

REMARK. This appendix is based on the result in [10]. A variant of this method is discussed in [8]. Also in [2] a plate problem is treated along the line of Johnson's method. Note that inequality (3) is exactly equal to the following relations.

$$D_t\varepsilon^n - CD_t\sigma^n = 0, \quad D_t\alpha^n = 0 \quad \text{if } f(\sigma^{n+1} - \alpha^n) \leq z_0$$

$$\begin{cases} D_t\varepsilon^n - CD_t\sigma^n = \frac{1}{\eta}\partial f|_{(\sigma^{n+1} - \alpha^n)} (f(\sigma^{n+1} - \alpha^n) - z_0)/\Delta t \\ D_t\alpha^n = (\sigma^{n+1} - \alpha^n)(1 - z_0/ f(\sigma^{n+1} - \alpha^n))/\Delta t \quad \text{if } f(\sigma^{n+1} - \alpha^n) > z_0. \end{cases}$$

To show this equivalence use the facts that the vectors $\alpha^{n+1} - \alpha^n$ and $\sigma^{n+1} - \alpha^{n+1}$ are parallel and that $\partial f_{n+1} = S(\sigma^{n+1} - \alpha^{n+1})/z_0$. Hence σ^{n+1} can be determined independently of α^{n+1} in practice.

REFERENCES

1. Adams R.A. : Sobolev Spaces, Academic Press, New York-San Francisco-London, 1975.
2. Brezzi F., Johnson C., Mercier B. : Analysis of a mixed finite element method for elasto-plastic plates, Mathematics of Computation, Vol. 31, No. 140, 1977, 809 - 817.
3. Ciarlet P.G. : The Finite Element Method for Elliptic Problems, North-Holland, Amsterdam-New York-Oxford, 1976.
4. Duvaut G., Lions J.L. : Les Inequations en Mecanique et en Physique, Dunnod, Paris, 1972.
5. Ekeland I., Temam R. : Convex Analysis and Variational Problems, North-Holland American Elsevier, Amsterdam-Oxford, New York.
6. Fujii H. : Finite element schemes - Stability and convergence, Advances in Computational Methods in Structural Mechanics and Design, The University of Alabama Press, 1972, 201-218.
7. Hlavaček I., Nečas J.: Mathematical Theory of Elastic and Elasto-plastic Bodies, Elsevier, Amsterdam, 1980.
8. Hlavaček I. : A finite element solution for plasticity with strain - hardening, R.A.I.R.O. Analyse numérique, Vol. 14, No. 4, 1980, 347-368.
9. Johnson C. : On plasticity with hardening, Journal of Mathematical Analysis and Applications, Vol. 62, 1978, 325-336.
10. Johnson C. : On finite element methods for plasticity problems, Numerische Mathematik, Vol. 26, 1976, 79-84.
11. Kachanov L.M. : Foundations of the Theory of Plasticity, North-Holland,

Amsterdam-London, 1971.

12. Lang S. : Analysis II, Addison-Wesley, 1969.

13. Miyoshi T. : Elastic-plastic vibration of a rod, Publications of R.I.M.S. Kyoto University, Vol. 16, No. 2, 1980, 377-392.

14. Miyoshi T. : On existence proof in plasticity theory, Kumamoto Journal of Science, Vol. 14, No. 1, 1980. 18-33.

15. Miyoshi T. : Numerical stability in dynamic elastic-plastic problems, R.A.I.R.O. Analyse numérique, Vol. 14, No. 2, 1980, 175-188.

16. Miyoshi T. : A note on the classical solutions of semi-discrete quasi-static plasticity problems, Kumamoto Journal of Science, Vol. 15, 1982 7-10.

17. Miyoshi T. : Yielding and unloading in semidiscrete problems of plasticity, Nonlinear Partial Differential Equations in Applied Science; Proceedings of the U.S.-Japan Seminar, Tokyo, 1982, H. Fujita, P.D. Lax, G. Strang (eds.), Kinokuniya/ North-Holland, Tokyo, 1983, 189-204.

18. Nitsche J. A. : On Korn's second inequality, R.A.I.R.O. Analyse numérique, Vol. 15, No.3, 1981, 237-248.

19. Yamada Y. : Plasticity - Visco • elasticity, Baifukan, Tokyo, 1972.

20. Ziegler H. : A modification of Prager's hardening rule, Quart. Appl. Math. Vol. 17, 1959, 55-65.

INDEX